D0858390

CAMBRIDGE TRACTS IN MATHEMATICS

69. Simple Noetherian rings

JOHN COZZENS
Department of Mathematics
Rider College

AND

CARL FAITH
Department of Mathematics
Rutgers University

Simple Noetherian rings

CAMBRIDGE UNIVERSITY PRESS
CAMBRIDGE
LONDON · NEW YORK · MELBOURNE

Published by the Syndics of the Cambridge University Press
The Pitt Building, Trumpington Street, Cambridge CB2 1RP
Bentley House, 200 Euston Road, London NW1 2DB
32 East 57th Street, New York, N.Y.10022
296 Beaconsfield Parade, Middle Park, Melbourne 3206, Australia

First published 1975

Typeset in Great Britain by William Clowes, Great Yarmouth

Printed in the United States of America

Library of Congress Cataloguing in Publication Data
Cozzens, John, 1942–
Simple noetherian rings.
(Cambridge tracts in mathematics; 69)
Bibliography: p. 129
Includes index.
1. Noetherian rings. I. Faith, Carl Clifton,
1927– joint author. II. Title. III. Series.
QA251.5.C69 512'.2 75-11535
ISBN 0 521 20734 7

This book is for
Midge and Mickey;
Barbara and Kathy;
Heidi and Cindy

Contents

Preface

This book is intended for the general reader of mathematics, and the authors have eliminated unnecessary mathematical machinery whenever possible, or postponed using it as long as possible. This was done, however, only after the authors saw that it could be done without any loss of clarity, or generality, in the statements of theorems.

The book is as elementary and self-contained as practicable; and the little background required in homological and categorical algebra is listed in two short appendices. Otherwise, full definitions are given, and short, elementary full proofs are supplied for such tool theorems as the Morita theorem 1.12 (p. 15), the correspondence theorem 1.13 (p. 17), the Wedderburn–Artin theorem 2.23 (p. 32), the Goldie–Lesieur–Croisot theorem 4.4 (p. 65), and many others.

The authors are indebted to Professor Hyman Bass, one of the general editors of this series, for suggesting that such a study of simple Noetherian rings would be of interest. We are immensely grateful to Professor Bass and to the Cambridge University Press for making it possible for this study to appear in the Cambridge Tracts in Mathematics.

J.H.C.
C.F.

List of symbols†

	Name	Example
iff	double implication (or equivalence of propositions)	A is true iff B is true
$\forall$	universal quantifier	$\forall a \in A$
$\exists$	existential quantifier	$\exists a \in A$
$\in$	membership	$a \in A$
$\notin$	nonmembership	$a \notin A$
$\subset$	proper containment	$A \subset B$
$\subseteq$	containment	$A \subseteq B$
$\Rightarrow$	implication	$A \Rightarrow B$
$\Leftrightarrow$	double implication	$A \Leftrightarrow B$
$=$	equals	$A = B$
$\varnothing$	empty set	—
$\mathbb{N}$	natural numbers	$1, 2, \ldots$
$\mathbb{Z}$	integers	$0, \pm 1, \pm 2, \ldots$
$\mathbb{Q}$	rational numbers	a/b, $a, b \in \mathbb{Z}$, $b \neq 0$
$\cup$	union	$A \cup B$
$\cap$	intersection	$A \cap B$
$+$	plus	$a + b$
$-$	minus	$a - b$
$\circ$	composition	$f \circ g$
$\times$	Cartesian product	$A \times B$
$\to$	mapping	$A \to B$
$f: A \to B$	mapping	—
$\mapsto$	mapping	$a \mapsto b$
$\hookrightarrow$	embedding	$A \hookrightarrow B$
$\times$	cross	$\times_i A_i$
$\prod$	product	$\prod_i A_i$
$\coprod$	coproduct	$\coprod_i A_i$

† Much of the material in this table is taken from *Grundlehren* volume 190 and is reproduced by permission of Springer-Verlag, Heidelberg.

Symbol	Meaning	Example
$\approx$	equivalence or isomorphism (in the appropriate category)	$A \approx B$, mod-$A \approx$ mod-B
$\simeq$	equivalence (as functors)	$S \simeq T$
$\overset{\text{nat}}{\approx}$	natural isomorphism (or equivalence)	$A \overset{\text{nat}}{\approx} B$
$>$	greater than	$a > b$
$<$	less than	$a < b$
$\geq$	greater than or equal to	$a \geq b$
$\leq$	less than or equal to	$a \leq b$
$\rightsquigarrow$	functor	$A \rightsquigarrow B$
∇	wedge	$A \nabla B$
$\perp$	perpendicular ("perp")	$A^{\perp}$ and ${}^{\perp}A$
$\sum$	summation	$\sum_{i \in I} A_i$
$\oplus$	direct sum	$A \oplus B$, $\sum_{i \in I} \oplus A_i$
$\otimes$	tensor product	$A \otimes B$
$\square$	end of proof (occasionally, no proof)	—
$\sim$	similarity (Morita equivalence)	$A \sim B$
$\overset{Q}{\sim}$	equivalence (as orders)	$A \overset{Q}{\sim} B$
$\overset{l}{\sim}$	left equivalence (as orders)	$A \overset{l}{\sim} B$
$\overset{r}{\sim}$	right equivalence (as orders)	$A \overset{r}{\sim} B$

mod-A = the category of (unital) right A-modules
A-mod = the category of left A-modules
Ab = the category of abelian groups (mod-$\mathbb{Z}$)
L.H.S. (R.H.S.) = left (right) hand side
$\{x \mid x \text{ has property } P\}$ = the set of all x with property P
$\{x_i \mid i \in I\}$ = a set of elements indexed by a set I

Conventions

The following conventions are the standard ones adopted by most modern texts on homological algebra or ring theory. Their inclusion is primarily for the reader's convenience.

For the most part, by a ring A, we mean an associative ring with unit, usually denoted 1. We let ring-1 denote a ring in which an identity element is not assumed. Thus, a ring-1 may have 1. All ring homomorphisms will be assumed to preserve the respective units. We let mod-A (A-mod) denote the category of all unital right A-modules (left A-modules), the morphisms being right (left) A-linear homomorphisms. The fact that M is a right (left) A-module will be denoted variously by $M \in$ mod-A ($M \in A$-mod) and M_A (${}_AM$).

$f: M_A \to N_A$ ($f: {}_AM \to {}_AN$) will indicate that f is a morphism (A-linear homomorphism) between right A-modules (left A-modules) M and N.

If $f: M_A \to N_A$ ($f: {}_AM \to {}_AN$), we shall write f on the left (right) of its argument. Thus $f(m)$ ($(m)f$) will denote the image of f on $m \in M_A$ ($m \in {}_AM$).

If M, $N \in$ mod-A and $\text{hom}_A(M_A, N_A) = \{f \mid f: M_A \to N_A\}$, we set End $M_A = \text{hom}_A(M_A, M_A)$. Similarly, End ${}_AM = \text{hom}_A({}_AM, {}_AM)$. Of course, End M_A (End ${}_AM$) is called the **endomorphism ring** of M_A (${}_AM$).

Whenever $M \in$ mod-A and $M \in B$-mod and $b(ma) = (bm)a$, $\forall m \in M$, $a \in A$, $b \in B$, we call M a (B, A)-bimodule and denote this variously by $M \in B$-mod-A or ${}_BM_A$. The morphisms in the category B-mod-A are homomorphisms $f: M \to N$ which satisfy

$$f(bma) = bf(m)a \quad \forall m \in M, a \in A, b \in B.$$

Let $M \in$ mod-A and $B =$ End M_A. Since we have agreed to write homomorphisms on the left of their arguments, M is naturally a left B-module, in fact a (B, A)-bimodule where

$b \cdot m = b(m)$, $\forall b \in B$, $m \in M$. Similarly, if $M \in A$-mod and $B = \text{End}\,{}_A M$, $M \in A$-mod-B.

If $M \in B$-mod and $A = \text{End}\,{}_B M$ then $\bar{B} = \text{End}\, M_A$ is called the **biendomorphism ring (bicommutator)** of M and is frequently denoted by $\bar{B} = \text{Biend}\,{}_B M$. Clearly, B maps homomorphically into $\bar{B}$ (as rings) via left multiplication. This map is obviously injective whenever ${}_B M$ is faithful. If the map is a surjection ${}_B M$ is called **balanced**. In this case, center $A \approx$ center B via the canonical map of B into $\text{Biend}\,{}_B M$.

A **basis** is another term for a generating set of a module. However, **free basis** means that a basis $\{x_i \mid i \in I\}$ is **linearly independent** in the sense that any (finite) linear combination $\sum_{i \in I} x_i r_i$ is zero iff $r_i = 0$ for every i. (Finite here means that $r_i = 0$ for all but a finite number of $i \in I$.) In this case, the sum $\sum_{i \in I} x_i R$ of the submodules generated by the set $\{x_i \mid i \in I\}$ is direct, notationally, $\sum_{i \in I} \oplus\, x_i R$.

As indicated in the list of symbols, $M = A \oplus B$ denotes a **direct sum** of modules A and B, and then A (also B) is said to be a **direct summand** of M, or more briefly, a **summand** of M. (Thus, every $m \in M$ can be expressed in one and only one way, $m = a + b$, as a sum of elements $a \in A$ and $b \in B$.)

In order to save space, in defining 'right-handed' concepts, we shall assume the left-handed concept defined by right-left symmetry. (Similarly for left-handed concepts.) Example: right Noetherian.

A mapping $f: X \to Y$ of sets X and Y is said to be **injective**, or an **injection** provided that $f(x) = f(y) \Rightarrow x = y$, $\forall x, y \in X$. It can be shown that a mapping $f: X \to Y$ is injective iff f is **left cancelable** in the category of set morphisms in the sense that for any two mappings $g: U \to X$ and $k: U \to X$, then $fg = fk$ iff $g = k$.

Dually, a mapping $f: X \to Y$ is **surjective**, or a **surjection**, iff $Y = \text{im}\, f$. This happens iff f is **right cancelable**.

The terms map and mapping are synonyms for morphism in the specified category. Thus, a homomorphism of groups is also called a map. We use **monomorphism**, or the abbreviation, **mono**, in the sense that the morphism is left cancelable in the category; dually for **epimorphism**, or **epi**. Thus, an injection is mono in the category SETS of sets, a surjection is epi, and conversely. Similarly, in Ab (the abbreviation for mod-$\mathbb{Z}$, the category of abelian groups), a morphism $f: X \to Y$ is mono iff $\ker f = 0$ iff f is injective, and dually, f is epi iff $\operatorname{cok} f = 0$ iff f is surjective. Here, of course, $\ker f$ denotes the kernel of f, and $\operatorname{cok} f$ denotes the cokernel (which is isomorphic to $Y/\operatorname{im} f$).

The term 'dually' referred to above is meant in the sense that one concept is obtained from the other by reversing arrows and directions in the definitions. For a more explicit statement of duality, consult Faith [73*a*, p. 60].

A sequence $X \to Y \to Z$ of modules is **exact** if the image of $X \to Y$ is the kernel of $Y \to Z$. This terminology also applies to long sequences of modules and maps, namely

$$\ldots \to X_i \to X_{i+1} \to X_{i+2} \to \ldots \to X_n \to X_{n+1} \to \ldots$$

is **exact**, provided that each $X_i \to X_{i+1} \to X_{i+2}$ is exact for every i, with possibly i ranging over $\mathbb{Z}$. An exact sequence

$$0 \to X \to Y \to Z \to 0$$

is said to be **short exact**. Thus, $0 \to X \to Y$ is exact iff $X \to Y$ is mono, and $X \to Y \to 0$ is exact iff $X \to Y$ is epi. Thus there is always a canonical exact sequence

$$0 \to X \to Y \to Y/X \to 0$$

for any submodule X of Y.

Other requirements from homological algebra will be administered *ad hoc*, or, in a few cases, found in the Appendix, §1 and §2. The term **natural isomorphism** is an old one which has been subsumed under the concept of natural equivalence of functors defined in the Appendix, §2, but often we use it in its old informal sense to mean that the mapping indicated by $\overset{\text{nat}}{\approx}$ is the canonical one, and is an isomorphism.

A module M of mod-A is **Noetherian** if it satisfies the equivalent conditions:

(N_1) **a.c.c.**: Any ascending sequence

$$M_1 \subseteq M_2 \subseteq \cdots \subseteq M_n \subseteq \cdots$$

of submodules is ultimately stationary, that is, there exists an integer $N > 0$ such that $M_N = M_{N+k}\ \forall k \geq 1$.

(N_2) M has the maximum condition for submodules, that is, any nonempty set S of submodules contains a maximal element $M(S)$. (Thus, $M(S) \not\subset M'$ for any $M' \in S$.)

(N_3) Every submodule of M has a finite basis, that is, is finitely generated.

Dual to the concept of Noetherian is the concept of **Artinian**, namely, a right A-module M is Artinian provided that M satisfies the equivalent conditions:

(A_1) **d.c.c.** Any descending sequence

$$M_1 \supseteq M_2 \supseteq \cdots \supseteq M_n \supseteq \cdots$$

of submodules is ultimately stationary.

(A_2) M satisfies the minimum condition for submodules.

The ring itself is **right Noetherian (Artinian)** provided that A is a Noetherian (Artinian) module in mod-A. This happens iff every finitely generated module in mod-A is Noetherian (Artinian), a fact which shows that this is a Morita invariant property (in the sense defined on p. 15.) Furthermore, a theorem of Hopkins and Levitzki states that any right Artinian ring A is right Noetherian. (Since this result for simple rings is an immediate consequence of the Wedderburn–Artin theorem 2.23, p. 32, we omit the proof.)

The **Hilbert Basis Theorem** states that if A is a right Noetherian ring, then so is the polynomial ring $A[X]$ over A. This readily generalizes to finitely many variables. (See, for example, Faith [73*a*], p. 341.)

A module N is **simple** if N has no nontrivial submodules. For $M \in$ mod-A, the **socle** of M, soc M, is the sum of all simple submodules of M. If soc $M = M$, we call M **semisimple**. 'Dually', a submodule N of M is a **maximal submodule** of M or simply maximal, if M/N is simple, and the **radical** of M, rad M, is the intersection of all maximal submodules of M.

An ideal P of A is called **prime (semiprime)** if for all ideals I and J of A, $IJ \subseteq P \Rightarrow I \subseteq P$ or $J \subseteq P$ ($I^2 \subseteq P \Rightarrow I \subseteq P$). Here and hereafter, by an ideal of A, we mean a two-sided ideal of the ring A. A ring A is **prime** (**semiprime**) if 0 is a prime (semiprime) ideal of A.

A ring A is **simple** provided that there are no ideals except the trivial ones, A and 0. Clearly, every simple ring is prime (0 is a maximal ideal!). Simplicity of a ring A may be characterized by the property that every right module $M \neq 0$ is faithful, a Morita invariant property which shows that any full $n \times n$ matrix ring A_n over a simple ring A is simple.

The term **domain** is an abbreviation for an integral domain, that is, a ring without zero divisors $\neq 0$. Finally, a **field**, defined as a ring in which every element $x \neq 0$ has a (two-sided) inverse x^{-1}, may be characterized by the property that the only right ideals are the trivial ones. Thus, a field is simple.

Introduction

In this book we have collected the known theorems on the structure of simple right Noetherian rings, and more generally for simple rings containing a uniform right ideal U. By the Goldie [58, 60]† and Lesieur–Croisot [59] theorems, any prime right Noetherian ring R has a simple Artinian ring $Q(R)$ of right quotients. By the Artin [27]–Wedderburn [08] theorem, $Q(R)$ is isomorphic to a full $n \times n$ matrix ring D_n over a (not necessarily commutative) field D. Moreover, the endomorphism ring B of any uniform right ideal U of R is a right Ore [31] domain with right quotient field $Q(B) \approx D$.

The Goldie and Lesieur–Croisot theorem is historically the first representation of a right Noetherian non-Artinian prime ring in a matrix ring of finite rank with entries in a field. (The Chevalley–Jacobson density theorem for a primitive ring R yields a representation by infinite matrices over a field unless R is Artinian.)

Significant simplifications in, and additions to, the Goldie and Lesieur–Croisot theory occur if R is assumed to be a simple ring. Henceforth R will denote a simple ring with identity element, and uniform right ideal U as described. We now outline the structure theory for R, which includes that for a simple right Noetherian ring.

1 *Endomorphism ring theorem* 2.17‡

Every simple ring R with uniform right ideal U is the endomorphism ring, $R \approx \text{End}\,_B U$ canonically, of a torsion free

† Goldie [58] refers to a work by A. W. Goldie published in 1958; co-authors are referred to in the text in hyphenated form, in conformity with the widespread practice in mathematics.

‡ The notation 2.17 refers to a theorem in Chapter 2 numbered 17. In general, for integers a and b, the notation a.b will denote a statement numbered b in Chapter a.

module ($=U$) of finite rank over a right Ore domain B (Faith [64]).

2 *Projective module theorem* 2.17

U is a finitely generated projective left B-module, where $B = \text{End}\, U_R$ (Hart [67]).

3 *Uniqueness of orders theorem* 2.22

Assume that $R \approx \text{End}\,_{B'}U'$ for a finitely generated projective module U' over a right Ore domain B'. Then there is an embedding $B' \to D$ such that B' and B are equivalent orders of D. (Equivalence of orders is defined in the sense of Jacobson [43] in Chapter 2 (p. 26).)

In the next theorem, $\text{trace}\,_{B'}U' = \sum_{f \in \text{Hom}_{B'}(U', B')} (U')f$, that is, the B'-submodule generated by all elements $(u)f$, with $u \in U'$, $f \in \text{Hom}_{B'}(U', B')$.

4 *Least ideal theorem* 2.6

In any representation $R \approx \text{End}\,_{B'}U'$ (as in 3), $T' = \text{trace}\,_{B'}U'$ is a (nonzero) least ideal of B'.

Thus, $T' = B'$ iff B' is a simple ring.

5 *Idempotent right ideal theorem* 2.21

Moreover, R is right Noetherian iff B' (as in 4) satisfies the ascending chain condition (a.c.c.) on idempotent right ideals contained in T'. In this case B satisfies the a.c.c. on principal right ideals.

6 *Single ideal theorem* 2.7

There is a module U' and domain B' (as in 3) such that $R \approx \text{End}\,_{B'}U'$ and such that B' contains at most one nontrivial ideal T'.

7 *Same centers theorem* 2.7

The center C of B is a field canonically isomorphic to the center of R, and U' and B' (in 6) can be chosen such that $U' = U$ and $B' = C + T$, where $T = \text{trace}\,_B U$.

In other words, U is finitely generated projective over $B' = C + T$, and $R \approx \text{End}\,_{B'} U$ canonically. Moreover, center $B' = C$ and trace $_{B'} U = T$.

8 *Hereditary ring theorem* 2.25

Every representation of R as an endomorphism ring $R \approx \text{End}\,_{B'} U'$ with $B' = \text{End}\, U'_R$ (as in 3) has the property that B' is a simple ring iff R is right hereditary.

This shows that the single ideal theorem is of some use: the representation of R over a domain with a single nontrivial ideal does occur (as in 6) when R is a nonhereditary simple ring.

The results 3 to 8 are proved by Faith [72*a*] and are consequences of the correspondence theorem for projective modules taken up in Chapter 1.

Any simple Noetherian ring R is a maximal order in $Q(R)$, in a sense defined in Chapter 2 (p. 27), and the question arises when is $B = \text{End}\, U_R$ a maximal order in D. When R is a two-sided order we have:

9 *Maximal order theorem* 4.14

The domain $B = \text{End}\, U_R$ is a maximal order in D iff U is a reflexive right R-module. This happens, for example, when U is a maximal uniform (=basic) right ideal.

By a theorem of Goldie, any uniform right ideal Y of R is contained in a basic right ideal. This, together with the previous theorems shows that R can be represented as the endomorphism ring of a finitely generated projective over a maximal order in a field.

10 *Reflexive ideal theorem* 4.20

If R is two-sided Noetherian with right global dimension $R \leq 2$, then every uniform right ideal of R is reflexive iff R is right hereditary.

Theorems 9 and 10 are proved by Cozzens [75], and taken together, Theorems 1 to 10 reduce many structural questions on a simple right Noetherian ring R to corresponding questions on the structure of a maximal right order B (of a field D) having at most one nontrivial ideal. For example, the question of when U can be chosen such that $B = \text{End } U_R$ is itself a simple ring can be formulated entirely within D, namely

11 *Simple endomorphism rings theorem* 2.20

Assume that R has the representation $R = \text{End }_{B'}U'$ as in 3. Then, there exists a simple right Ore domain B_0 and a finitely generated projective left B_0-module U_0 such that $R \approx \text{End }_{B_0}U_0$ iff there exists a right ideal I of B' such that

$$(I : I) = \{d \in D \mid dI \subseteq I\}$$

is a simple subring of D. (Note that since $\text{End } I_{B'} \approx (I : I)$, this is equivalent to requiring that some right ideal of B' has simple endomorphism ring.) (Faith [72*a*].)

The question of the existence of such a representation of R (as in 11) is one of the main structural problems on a simple right Noetherian ring R, and in this case (and only then) R is similar (= Morita equivalent) to a right Ore domain A, that is, then (and only then) there is a category equivalence mod-$R \approx$ mod-A, for some domain A. A positive result in this direction is the following:

12 *Global dimension* 2 *theorem* 2.40

A simple Noetherian ring R of right global dimension ≤ 2 is similar to a (simple) domain A. (Faith [72*a*], Hart–Robson [70], Michler [69]). The proof requires ideas behind Bass's global dimension 2 theorem (4.16).

This holds if R is a principal right ideal ring, since then r.gl.dim $R \leq 1$, but in this case much more is true:

13 *Principal right ideal theorem* 4.7

Any prime principal right ideal ring R is isomorphic to a full $n \times n$ matrix ring A_n over a right Noetherian hereditary domain A. (Goldie [62].)

Curiously, A need not be a principal right ideal domain for this to happen (Swan [62], see also 3.6, p. 49). (When R is simple then of course A is.)

14 *Ore domains and the Faith–Utumi theorem* 4.6

A refinement of the Goldie–Lesieur–Croisot theorem is the theorem of Faith–Utumi (proved in Chapter 4) which states that if R is any right order in D_n, then R contains a right order F_n, where F is a right order in D. A change of matrix units in D_n may be necessary for this (unless R is also a left order) and, moreover, F does not in general contain an identity element. [If it did, then R would contain the full set of matrix units of D_n, and hence R would itself be a full $n \times n$ matrix ring H_n, where $H = R \cap D$. This is, in general, not the case.] One can show that F can be chosen to contain no nontrivial ideals when R is simple.

15 *Classification of simple domains*

Inasmuch as Ore domains play a role in the structure of simple Noetherian rings analogous to the role of fields in the structure of simple Artinian rings, their classification, especially the simple ones, is fundamental to the structure theory of simple Noetherian rings. The rest of the introduction is devoted to this.

Restricting our attention for the moment to simple integral domains, we can distinguish three main types:

(D_1) For any simple domain (or ring) R of characteristic 0, and finitely many commuting outer derivations $\delta_1, \ldots, \delta_n$, the ring $\mathscr{D}_R$ of differential polynomials is a simple right Noetherian Ore domain. (Ore [33], Littlewood [33], Amitsur [57].) We call these **Amitsur–Littlewood** (simple Ore) **domains.**

(D_2) For any field F, the localization R_M at a maximal ideal M, of the skew polynomial ring $R = F[x]$ with respect to a ρ-derivation δ is a simple principal left ideal (pli-) domain, not a field. (Jacobson [43].) We call these domains **simple localizations.**

(D_3) For any field F, if $F[x]$ is a primitive ring, e.g., if F is a transcendental algebra over the center C, then $A = F \otimes_C C(x)$ is a simple pli and pri (principal right ideal)-domain.

D_3 is an observation derivable from Jacobson's theorem [64*a*] characterizing when $F[x]$ is a primitive ring. (See Theorem 3.20 and 3.21, pp. 61–2.) For this reason, we entitle this class **Jacobson domains.** (Actually, D_3 is a subclass of D_2.)

16 *V-domains*

In this the concluding section of the Introduction, we summarize some results on a special class of simple Ore domains for which there is a rather satisfactory structure theory paradoxically resembling (commutative) Dedekind domains.

The question of the existence of integral domains, not fields, having the property that every simple right module is injective was first raised by Faith [67*a*, Problem 17, p. 130]. The history of this problem is as follows: Kaplansky showed that commutative regular rings could be characterized by the injectivity of the simple modules. We use the terminology right K-ring (after Kaplansky) for a ring with every simple right module injective. Villamayor characterized a right K-ring R by the property that every right ideal is the intersection of maximal right ideals, or equivalently, every right module M has the property that the intersection of its maximal submodules, namely rad M, is equal to 0. Such rings are called right V-rings, and we refer interchangeably to right K-rings and V-rings.

The first examples of right V-domains, not fields, given by Cozzens [70], were differential polynomial rings $K[x, d]$ with respect to a derivation d over a (Kolchin) universal differential field K. Analogous examples of these right V-domains were localizations of twisted polynomial rings $K[x, \sigma]$ with respect to an automorphism σ of a field K over which certain designated algebraic equations have solutions. These are taken up in Chapter 5.

More generally, simple Noetherian V-domains, not fields, were shown to exist in all homological dimensions by Cozzens–Johnson [72]. Osofsky has shown the existence of V-domains having infinitely many nonisomorphic simple modules (see 6.20). However, the problem of exhibiting examples with just finitely many nonisomorphic simple modules remains open.

All of the above examples which are pri-domains also have the property that each cyclic module, not isomorphic to the base ring, is injective. Such rings are called PCI-rings (proper cyclics injective). Chapter 6 is devoted to the study of these rings: they are either semisimple Artinian, or else semihereditary simple right Ore domains in which any finitely generated right ideal I can be generated by two elements, one of which is an arbitrary nonzero element of I. Other resemblances to Dedekind domains are noted.

1. *The correspondence theorem for projective modules*

In this chapter, we prove the correspondence theorem for projective modules, referred to in the Introduction, on which most of the known structure theory for simple Noetherian rings depends.

If $M, N \in$ mod-A and $B = \text{End}\, M_A$, then **the trace of N in M** is the A-submodule of M generated by all elements $\{f(x) \mid x \in N, f \in \hom_A(N, M)\}$. This module is denoted trace ${}_M N$ and is a (B, A)-bimodule. In particular, trace ${}_A M$ is an ideal of A for any module M.

1.1 PROPOSITION AND DEFINITION *An object U of* mod-A *is a* **generator** *if the equivalent conditions hold:*

(1) *The group-valued functor* $\hom_A(U, -)$: mod-$A \rightsquigarrow$ Ab (*the category of abelian groups*) *is faithful. Thus, given any nonzero $k : X_A \to Y_A$ there exists a homomorphism $h : U_A \to X_A$ such that* $kh \neq 0$.

(2) trace ${}_M U = M$ *for all* $M \in$ mod-A.

(3) trace ${}_A U = A$.

(4) $U^n \approx A \oplus X$ *for some integer* $n > 0$ *and* $X \in$ mod-A.

(5) $U^n \to A \to 0$ *is exact for some* $n > 0$.

Proof For any index set I, $U^{(I)}$ will always denote the direct sum (coproduct) of I copies of U.

(1) $\Rightarrow$ (2): Define

$$\pi \begin{cases} U^{(\hom_A(U,M))} \to M, \\ (\ldots, u_f, \ldots) \mapsto f(u) \quad \forall f \in \hom_A(U, M). \end{cases}$$

Note that $\text{im}\, \pi = \text{trace}\, {}_M U$. If $\text{im}\, \pi \neq M$, set $C = M/\text{im}\, \pi$ and let $\nu : M \to C$ be the canonical projection. By assumption, there exists an $f \in \hom_A(U, M)$ such that $\nu f \neq 0$. However, $\text{im}\, f \subseteq \text{im}\, \pi$, contradicting $\nu f \neq 0$. Thus, $\text{im}\, \pi = M$.

(2) $\Rightarrow$ (1): Let $g : M \to N$ be nonzero. If $gf = 0$ for all $f \in \hom_A(U, M)$, then clearly $g\pi = 0$ implying that $g = 0$ by surjectivity of π. Thus, $gf \neq 0$ for some $f \in \hom_A(U, M)$.

(2) $\Rightarrow$ (3) is clear; the equivalence of (3), (4) and (5) is immediate by 1.2 which follows.

(3) $\Rightarrow$ (1): Since A is free, there exists an epi $A^{(I)} \rightarrow M$ for some set I and hence, an epi $U^{(I)} \rightarrow M$. By the same arguments used in the proof (2) $\Rightarrow$ (1), (1) is clearly satisfied. □

1.2 PROPOSITION AND DEFINITION *An object* $U \in$ B-mod *is* **projective** *if the following equivalent conditions hold:*

(1) *The group valued functor* $\hom_B(U, -) : B\text{-mod} \rightsquigarrow \text{Ab}$ *is exact.*

(2) (Dual basis lemma) *There exist sets* $\{x_i \mid i \in I\}$ *of elements of* U *and* $\{f_i \mid i \in I\}$ *of elements of* $\hom_B(U, B)$ *such that for each* $x \in U$, $(x)f_i = 0$ *for almost all* $i \in I$ *and*

$$x = \sum_{i \in I} (x)f_i x_i \quad \forall x \in U.$$

(3) U *is isomorphic to a (direct) summand of a free* B*-module.*

(4) *Every exact sequence* $M \rightarrow U \rightarrow 0$ *in* B-mod *splits.*

Proof The equivalence of (1), (3) and (4) is elementary, and will be left for the reader. We shall contend with the equivalence of (2) and (3).

(4) $\Rightarrow$ (2): Let $\{x_i \mid i \in I\}$ be *any* generating set for ${}_BU$, $\{e_i \mid i \in I\}$ a basis for $B^{(I)}$ and $f : B^{(I)} \rightarrow U$ the B-map defined by $e_i \mapsto x_i$. Since f splits there exists a map $g : U \rightarrow B^{(I)}$ such that $gf = 1_U$.

For any $x \in U$,

$$(x)g = \sum_{i \in I} b_i e_i,$$

where $b_i = 0$ for almost every $i \in I$. Define a family $\{f_i \mid i \in I\}$, where each $f_i : {}_BU \rightarrow {}_BB$, by $x \mapsto b_i$ for all $x \in U$. Clearly,

$$\begin{aligned} x &= \sum b_i x_i \\ &= \sum (x)f_i x_i \qquad \forall x \in U. \end{aligned}$$

(2) $\Rightarrow$ (3): Let $f : B^{(I)} \rightarrow U$ be defined as before and define $g : U \rightarrow B^{(I)}$ by

$$(u)g = \sum_{i \in I} (u)f_i e_i \quad \forall u \in U.$$

Clearly, g is a B-map and $gf = 1_U$. □

Remarks (a) The proof shows card I can be chosen equal to the cardinal of any basis of U, and, moreover, the equation for x in (2) shows that $\{x_i \mid i \in I\}$ is a basis of U.

(b) The set $\{f_i \mid i \in I\}$ is a basis of the 'dual' module $\text{Hom}_B(U, B)$ if $_BU$ is finitely generated.

(c) If A is any simple ring, then any right ideal $U \neq 0$ is a generator of mod-A, for $T = \text{trace}\,_AU$ is an ideal containing AU, hence $T = A$.

1.3 PROPOSITION *If U is projective in B-mod, then*

$$T = \text{trace}\,_BU = \bigcap_{\substack{K \subseteq B \\ \text{and } KU = U}} K.$$

That is, T is the intersection of all right ideals K of B such that $KU = U$. In particular, $U = TU$. Moreover, $T^2 = T$.

Proof By (2) of 1.2, $U = TU$ since each $(u)f_i \in T$. Thus, the L.H.S. $\supseteq$ R.H.S. Conversely, if $U = IU$, for some (right) ideal of B, then

$$T = \text{trace } U = \text{trace}(IU) = IT \subseteq I.$$

Thus, R.H.S. $\supseteq$ L.H.S. The fact that $T^2 = T$ follows immediately from the dual basis lemma. $\square$

1.4 CUT-DOWN PROPOSITION *Let U be projective*

$$B^{(I)} \xrightarrow{p} U \to 0$$

exact in B-mod, *and let* $T = \text{trace}\,_BU$. *If B_0 is any subring of B containing T, then U is projective in B_0*-mod *and p induces an exact sequence*

$$B_0^{(I)} \to U \to 0.$$

Furthermore, the inclusion map $\text{End}\,_BU \to \text{End}\,_{B_0}U$ *is an isomorphism. If U is faithful over B, then*

$$\text{center } B_0 = B_0 \cap \text{center } B.$$

Proof Let $T^{(I)} \subseteq B_0^{(I)}$ be the canonical inclusion. Then,

$$p(B_0^{(I)}) \supseteq p(T^{(I)}) = Tp(B^{(I)}) = TU = U$$

that is, $p \mid B_0^{(I)}$ is surjective. Since $_BU$ is projective, there exists a map $s : U \to B^{(I)}$ satisfying $sf = 1_U$. Clearly, im $s \subseteq$

$B_0^{(I)}$ and therefore, $B_0^{(I)} \to U \to 0$ is split exact. Thus, U is projective in B_0-mod.

Now let $t \in T$, $u \in U$, $b \in B$ and let $f \in \operatorname{End}_{B_0} U$. Since $bt \in T \subseteq B_0$,

$$(btu)f = bt(u)f = b(tu)f.$$

Since $U = TU$ is generated by the set $\{tu \mid t \in T, u \in U\}$, it follows that $f \in \operatorname{End}_B U$. This proves that $\operatorname{End}_B U = \operatorname{End}_{B_0} U$.

Finally, if $x \in$ center B_0, $b \in B$ and $t \in T$, then

$$(xb)t = x(bt) = (bt)x = b(tx) = b(xt) = (bx)t.$$

Hence, $(xb - bx)T = 0$ which implies that $(xb - bx)U = 0$ since $TU = U$. Since U is faithful, $xb = bx$. Thus, $x \in$ center B. □

1.5 PROPOSITION (Schanuel's Lemma) *Let*

$$0 \to K_i \to P_i \xrightarrow{\pi_i} M \to 0, \quad i = 1, 2$$

be two resolutions for M_A with P_i projective in mod-A, $i = 1, 2$. *Then $K_1 \oplus P_2 \approx K_2 \oplus P_1$.*

Proof Let P be the fiber product of π_1 and π_2, i.e., $P = \{(p_1, p_2) \in P_1 \oplus P_2 \mid \pi_1(p_1) = \pi_2(p_2)\}$. It is an easy matter to verify that the following diagram is commutative with exact rows and columns:

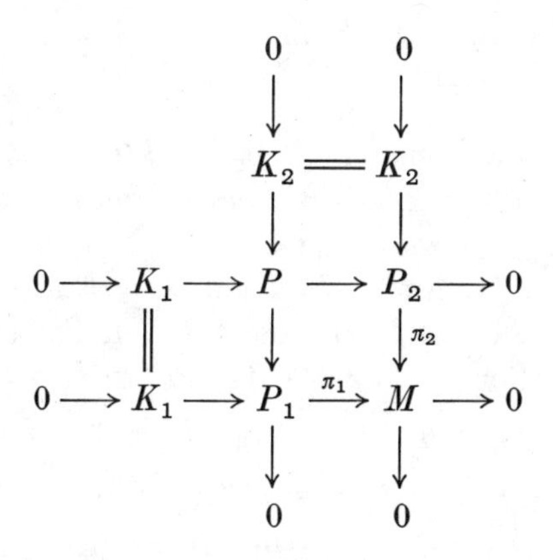

By projectivity of $P_1(P_2)$, $P \approx P_1 \oplus K_2$ $(P \approx P_2 \oplus K_1)$. □

1.6 Lemma (1) *Let* $U \in B$-mod *and* $A = \operatorname{End}_B U$. *The map* $\phi : \hom_B(U, B) \to \hom_A(U, A)$ *defined by*

$$[v](\phi(f)(u)) = ((v)f)\cdot u \quad \forall u, v \in U, f \in \hom_B(U, B)$$

is a bimodule homomorphism.

(2) *By symmetry, if* $B = \operatorname{End} U_A$, *the map* $\psi : \hom_A(U, A) \to \hom_B(U, B)$ *defined by*

$$((u)\psi(f))[v] = u\cdot f(v) \quad \forall u, v \in U, f \in \hom_A(U, A)$$

is a bimodule homomorphism.

In this case, ϕ *is an isomorphism with inverse* ψ.

Proof The proofs of (1) and the first part of (2) are routine.

Temporarily denote $\phi(f)$ by $[f,\ ]$ and $\psi(g)$ by $(\ ,g)$. Then

$$[v]([f,\ ](u)) = (v)f\cdot u \quad \forall u, v \in U, f \in \hom_B(U, B).$$

Let $f \in \hom_B(U, B)$. Then

$$f \overset{\phi}{\mapsto} [f,\ ] \overset{\psi}{\mapsto} (\ ,[f,\ ])$$

and

$$((u)(\ ,[f,\ ]))[v] = u\cdot[f, v] = (u)f\cdot v \qquad \forall u, v \in U, f \in \hom_B(U, B),$$

since $[f, v] \in A$ and the right A action on U is evalution. Thus, $\psi \circ \phi = 1$ on $\hom_B(U, B)$. Similarly, $\phi \circ \psi = 1$. □

1.7 Proposition *Let* $V \in B_0$-mod *be faithful and* $B = \operatorname{Biend}_{B_0} V$. *If* $U_0 = \hom_{B_0}(V, B_0)$ *and* $U = \hom_B(V, B)$, *the canonical embedding* $B_0 \hookrightarrow B$ *induces an inclusion* $U_0 \subseteq U$. *Moreover, if* $T_0 = \operatorname{trace}_{B_0} V$, *then* $BT_0 \subseteq B_0$.

Proof Let $A = \operatorname{End}_{B_0} V$. Then $B = \operatorname{End} V_A$. Let $(\ ,\ ) : \hom_{B_0}(V, B_0) \to \hom_A(V, A)$ be the (A, B_0)-homomorphism defined in 1.6. For $b \in B$, $v, v' \in V$ and $f \in U_0$, we have

$$\begin{aligned}(bv)f\cdot v' &= (bv)(f, v')\\ &= b((v)(f, v'))\\ &= b((v)f\cdot v')\\ &= (b(v)f)\cdot v'\end{aligned}$$

since $V \in B$-mod-A. Since ${}_BV$ is clearly faithful, $(bv)f = b(v)f$, $\forall b \in B$, $v \in V$ and $f \in U_0$. Thus, $U_0 \subseteq U$ as asserted. □

1.8 COROLLARY *Let U be a left B-module. Then, ${}_BU$ is a generator iff ${}_BU$ is balanced, and finitely generated projective over $A = \text{End }{}_BU$.*

Proof ⇒: That ${}_BU$ is balanced follows from the Proposition since $T = B$. Next

$$U^n \approx B \oplus X \quad \text{for some} \quad X \in B\text{-mod}$$

by projectivity of ${}_BB$. Apply $\text{hom}_B(-, U)$ and get the mod-A isomorphism

$$A^n \approx U \oplus \text{hom}_B(X, U).$$

Thus, U_A is finitely generated projective. The converse is obtained by reversing the steps. □

1.9 COROLLARY (1) *If U is a generator in* mod-A, *then*

$$U^* = \text{hom}_A(U, A) \approx \text{hom}_B(U, B)$$

is a generator in A-mod, *and finitely generated projective in* mod-B. *Moreover there is a canonical isomorphism $A \approx$* End U^*_B.

(2) *If U is finitely generated and projective in B*-mod, *then U is a generator in* mod-End ${}_BU$.

Proof Let

$$\langle\ ,\ \rangle \begin{cases} U \to U^{**} \\ u \mapsto \langle\ ,u\rangle \end{cases}$$

denote the canonical (B, A)-homomorphism of U into its A-bidual

$$U^{**} = \text{hom}_A(\text{hom}_A(U, A), A).$$

Thus, $f\langle\ ,u\rangle = \langle f, u\rangle = f(u)$, $\forall f \in U^*$, $u \in U$. If U is a generator in mod-A, then by 1.1, $1 = \sum_{i=1}^{n} f_i(u_i)$ for certain $f_i \in \text{hom}_A(U, A)$, $u_i \in U$. Consequently,

$$\sum_{i=1}^{n} f_i\langle\ ,u_i\rangle = 1_A$$

which shows that ${}_AU^*$ is indeed a generator. Then Corollary 1. implies the rest. □

1.10 Proposition and Definition *An object U of* mod-A *is a* **progenerator** *if U satisfies any of the following equivalent properties:*

(1) *U is a generator in* mod-A *and finitely generated and projective in* mod-A.

(2) *U is a generator in* mod-A *and in* End-U_A-mod.

(3) *U is finitely generated, balanced and projective both in* mod-A *and in* End U_A-mod.

(4) *U is a generator in* End U_A-mod, *finitely generated and projective in* End U_A-mod *and $A \approx$* Biend U_A *canonically.*

Proof An easy application of 1.8 and 1.9(2). □

1.11 Proposition and Definition *A bimodule ${}_BU_A$ is a (B,A)*-**progenerator** *provided the equivalent conditions hold:*

(1) *U is a progenerator in* mod-A *and $B \approx$* End U_A *canonically.*

(2) *U is a progenerator in* B-mod *and $A \approx$* End ${}_BU$ *canonically.*

(3) $\hom_A(U, -)$: mod-$A \rightsquigarrow$ mod-B *defines a category equivalence.*

Proof (1) and (2) are clearly equivalent by 1.10.

(3) ⇒ (1): Any equivalence is clearly faithful and exact, hence U_A is a progenerator by 1.1 and 1.2. Set T equal to the inverse of $S = \hom_A(U, -)$. Then $T \circ S \simeq 1_{\text{mod-}A}$ and $S \circ T \simeq 1_{\text{mod-}B}$. Now,

$$\begin{aligned} S(M) &\overset{\text{nat}}{\approx} \hom_B(B, S(M)) \\ &\overset{\text{nat}}{\approx} \hom_A(T(B), TS(M)) \\ &\overset{\text{nat}}{\approx} \hom_A(T(B), M) \end{aligned}$$

for all $M \in$ mod-A. In particular, $S(A) = \hom_A(U, A) \approx \hom_A(T(B), A)$ implying that $U_A \approx T(B)$. Hence,

$$B \approx ST(B) \approx S(U).$$

Thus, $B \approx$ End U_A.

(1) ⇒ (3): Let $S = \hom_A(U, -)$ and $T = \hom_B(U^*, -)$ where $U^* = \hom_A(U, A) \approx \hom_B(U, B)$ by 1.7.

$$\begin{aligned} TS(M) &= T(\hom_A(U, M)) \\ &= \hom_B(U^*, \hom_A(U, M)) \\ &\overset{\text{nat}}{\approx} \hom_A(U^* \otimes {}_BU, M) \\ &\overset{\text{nat}}{\approx} \hom_A(A, M) \overset{\text{nat}}{\approx} M. \end{aligned}$$

(The verifications that $U^* \otimes_B U \approx A$ and $U \otimes_A U^* \approx B$ will be deferred until after the proof of 1.12.)

Similarly, $ST(N) \overset{\text{nat}}{\approx} N$ for all $N \in$ mod-B. Thus, S is an equivalence with inverse T. □

1.12 PROPOSITION AND DEFINITION *Two rings A and B are* **similar** *or* **Morita equivalent**, $A \sim B$, *if they satisfy the equivalent conditions:*

(S_1) mod-$A \approx$ mod-B,

(S_2) A-mod $\approx$ B-mod,

(S_3) *There is a (B,A)-progenerator V.*

If $A \sim B$, then the lattice of right ideals of B is isomorphic to the lattice of A submodules of V under the map

$$I \mapsto IV.$$

Moreover, the lattice of ideals of A is isomorphic to the lattice of ideals of B.

Proof (S_1) and (S_3) are equivalent by 1.11 and (S_2) and (S_3) are equivalent by symmetry. The remaining statements follow immediately from the correspondence theorem 1.13 and will be omitted. □

If P is any property of a ring A or mod-A and whenever $A \sim B$, B has property P, P is said to be a **Morita invariant (property) of** A. For example, simplicity, r.gl.dim $R = n$, Artinian, Noetherian, semiprime Goldie are all Morita invariants while field, domain, and pli-domain are not. Of interest to us will be the weaker condition: if $A = \mathrm{End}\,_B V$, with $_B V$ finitely generated and projective, and A has property P, then so does B.

To that end, suppose $A = \mathrm{End}\,_B V$, where $_B V$ is finitely generated and projective. What are the relationships between the various module lattices which naturally arise? This question is answered completely in the correspondence theorem 1.13. In order to facilitate the proof of 1.13 we shall first make some definitions and observations.

Let $V \in B$-mod and $A = \mathrm{End}\,_B V$. Set $U = \hom_B(V, B)$. Then V and U are canonical (B, A) and (A, B)-bimodules respectively. Define two maps (,) and [,] as follows:

$$[\ ,\ ]: V \times U \to B,$$

where $(v, f) \mapsto [v, f] = (v)f$, $\forall v \in V, f \in U$.

$$(\, , \,) : U \times V \to A,$$

where (g, v') is that element of $A = \operatorname{End}_B V$ taking v to $(v)g \cdot v'$, that is,

$$[v](g, v') = (v)g \cdot v' \quad \forall v, v' \in V, g \in U.$$

One readily checks that $[\, , \,]$ and $(\, , \,)$ are balanced bilinear maps (balanced bilinear in the sense that, for example, the mappings $V \to B$ and $U \to B$ induced by $[\, , \,]$ are homomorphisms, and $[va, u] = [v, au]$, $\forall v \in V$, $u \in U$, $a \in A$) such that

$$v \cdot (u, v') = [v, u] \cdot v'$$

and

$$(u, v) \cdot u' = u \cdot [v, u'] \quad \forall u, u' \in U, v, v' \in V.$$

Since $[\, , \,]$ (resp. $(\, , \,)$) is a balanced bilinear map from $V \times U \to B$ (resp. $U \times V \to A$), $[\, , \,]$ (resp. $(\, , \,)$) induces a map, also denoted $[\, , \,]$ (resp. $(\, , \,)$),

$$[\, , \,] : V \otimes_A U \to B,$$
$$(\, , \,) : U \otimes_B V \to A.$$

Clearly, $[\, , \,]$ is a (B, B) homomorphism (resp. $(\, , \,)$ is an (A, A) homomorphism) and $\operatorname{im}[\, , \,] = \operatorname{trace}_B V$. Moreover, $[\, , \,]$ (resp. $(\, , \,)$) is epi if and only if it is an isomorphism. For suppose $[\, , \,]$ is epi and $x = \sum_i v_i' \otimes u_i'$ with $x \in \ker[\, , \,]$, that is, $\sum_i [v_i', u_i'] = 0$. Since $1 \in B$ can be written in the form

$$1 = \sum_j [v_j, u_j],$$

$$\begin{aligned} x &= \sum_j [v_j, u_j]\Big(\sum_i v_i' \otimes u_i'\Big) \\ &= \sum_{i,j} [v_j, u_j] v_i' \otimes u_i' \\ &= \sum_{i,j} v_j(u_j, v_i') \otimes u_i' \\ &= \sum_{i,j} v_j \otimes (u_j, v_i') u_i' \\ &= \sum_j v_j \otimes u_j \Big(\sum_i [v_i', u_i']\Big) = 0. \end{aligned}$$

Thus, $[\ ,\ ]$ is an isomorphism. The proof of the parenthetical statement is analogous. In fact, by 1.2(2), $(\ ,\ )$ is an isomorphism if and only if ${}_BV$ is finitely generated and projective.

For subsets $V_1 \subseteq V$, $U_1 \subseteq U$, set

$$[V_1, U_1] = \{[v_1, u_1] \mid v_1 \in V_1, u_1 \in U_1\}.$$

If $V_1 \in \text{mod-}A$ and $U_1 \in A\text{-mod-}B$, then clearly, $[V_1, U_1]$ is a right ideal of B. Similar definitions are made for (U', V') etc.

Set $T = \text{trace}\ {}_BV = [V, U]$. The important lattices which arise are

$$\begin{aligned} \text{lat}(B_B) &= \text{the lattice of right ideals of } B \\ \text{lat}(V_A) &= \text{the lattice of } A\text{-submodules of } V \\ \text{lat}({}_BV_A) &= \text{the lattice of } (B, A)\text{-submodules of } V \\ (\text{lat}(B_B))T &= \text{the lattice of right ideals of } B \text{ of the} \\ &\qquad \text{form } I' = IT, I \in \text{lat}(B_B). \end{aligned}$$

The lattice operations $\vee$ and $\wedge$ are defined on $(\text{lat } B_B)T$ by

$$I \vee J = I + J$$

and

$$I \wedge J = (I \cap J)T.$$

We also have $(\text{lat }{}_BB_B)T$ consisting of the set of all ideals of B of the form $I' = IT$, $I \in \text{lat }{}_BB_B$. (The operations $\vee$ and $\wedge$ are defined as in $(\text{lat } B_B)T$.) Others are $T(\text{lat }{}_BB_B)T$, and $T(\text{lat }{}_BV_A)$ consisting of the set of all (B, A)-submodules of V of the form $W' = TW$, $W \in \text{lat }{}_BV_A$. Here the operations $\vee$ and $\wedge$ are defined by

$$\begin{aligned} W_1 \vee W_2 &= W_1 + W_2, \\ W_1 \wedge W_2 &= T(W_1 \cap W_2). \end{aligned}$$

1.13 The Correspondence Theorem for Projective Modules. *Let* $A = \text{End }{}_BV$, ${}_BV$ *finitely generated projective and* $T = \text{trace }{}_BV$. *Then,*

(1) *The mapping*

$$\rho \begin{cases} \text{lat } V_A \to (\text{lat } B_B)T \\ \qquad W \mapsto [W, U] \end{cases}$$

is a lattice isomorphism with inverse defined by $K \mapsto KV$, *for any right ideal* $K = KT$ *of* B.

(2) *The mapping*

$$\begin{cases} \text{lat}\ {}_B V_A \to (\text{lat}\ {}_B B_B)T \\ W \mapsto [W, U] \end{cases}$$

induced by (1) *is a lattice isomorphism.*

(3) *The mapping*

$$\begin{aligned} \psi : \text{lat}\ {}_A A_A &\to T(\text{lat}\ {}_B V_A) \\ J &\mapsto VJ \end{aligned}$$

is a lattice isomorphism with inverse

$$W \mapsto (U, W).$$

(4) *The mapping*

$$\rho \circ \psi \begin{cases} \text{lat}\ {}_A A_A \to T(\text{lat}\ {}_B B_B)T \\ J \mapsto [VJ, U] \end{cases}$$

is an isomorphism of multiplicative semigroups.

Proof (1) Since $T = [V, U]$ and $A = (U, V)$,

$$\begin{aligned} [W, U]T &= [W, U][V, U] \\ &= [[W, U]\cdot V, U] \\ &= [W\cdot(U, V), U] \\ &= [W, U] \quad \forall W \in \text{lat}\ V_A. \end{aligned}$$

Thus, ρ is actually a mapping into (lat $B_B)T$.

Next, $[W_1 + W_2, U] = [W_1, U] + [W_2, U]$ and $[W_1 \cap W_2, U] = ([W_1, U] \cap [W_2, U])T$, $\forall W_1, W_2 \in \text{lat}\ V_A$ which together imply that ρ is a lattice homomorphism. The first statement is obvious. As for the second, since ${}_B V$ is projective,

$$\begin{aligned} ([W_1, U] \cap [W_2, U])\cdot V &= [W_1, U]\cdot V \cap [W_2, U]\cdot V \\ &= W_1 \cap W_2. \end{aligned}$$

Thus,

$$\begin{aligned} [W_1 \cap W_2, U] &= ([W_1, U] \cap [W_2, U])[V, U] \\ &= ([W_1, U] \cap [W_2, U])T \end{aligned}$$

as asserted.

Finally, the map is clearly an order homomorphism, as is the map

$$IT \mapsto ITV = IV$$

which is easily seen to be its inverse.

(2) Clear.

(3) ψ is clearly a map into $T(\text{lat}\,_B V_A)$ since $VI = (TV)I = T(VI)$. Similarly, the map $T(\text{lat}\,_B V_A) \to \text{lat}\,_A A_A$ defined by $TW \mapsto (U, TW)$ is a mapping into $\text{lat}\,_A A_A$.

Observe that

$$I \mapsto VI \mapsto (U, VI) = I$$

and

$$TW \mapsto (U, TW) \mapsto V\cdot(U, TW) = [V, U]TW = T^2W = TW.$$

Thus, the maps are mutually inverse to one another. That they are lattice homomorphisms follows from the fact that U_B is projective, and hence $(U, W_1 \cap W_2) = (U, W_1) \cap (U, W_2)$.

(4) Clear with the observation that

$$\begin{aligned}[VI, U][VJ, U] &= [[VI, U]VJ, U] \\ &= [VI\cdot(U, VJ), U] \\ &= [VIJ, U]\end{aligned}$$

$\forall I, J \in \text{lat}\,_A A_A$. □

1.14 Corollary *Let V be finitely generated faithful and projective over a ring B, let $A = \text{End}\,_B V$, and let $T = \text{trace}\,_B V$. For example, V can be an arbitrary generator in* mod-A, *for any ring A, and $B = \text{End}\, V_A$.*

(1) *Let P be any property of multiplicative semigroups which is preserved under homomorphisms. Then* (ideals of A) *has property P if and only if T*(ideals of B)*T has property P.*

(2) *A is semiprime if and only if $TQ = 0$ for every nilpotent ideal Q of B.*

(3) *A is prime if and only if $TQ = 0$ for every annihilator ideal Q of B.*

(4) *If B is semiprime, then B is prime if and only if A is prime.*

(5) *A has a finite ideal lattice if and only if the set of ideals of the*

form TIT for an ideal I of B is finite. In particular, A is a simple ring if and only if $TK = 0$ for every ideal $K \neq T$ contained in T.

(6) *If B is semiprime, then A is simple if and only if T is the least ideal of B. In this case a right ideal I of B satisfies $I = IT$ if and only if I is an idempotent right ideal contained in T.*

Proof The corollary is a consequence of the semigroup isomorphism

$$\text{ideals of } A \to T(\text{ideals of } B)T$$

stated in (4) of 1.13. Thus, if P is a statement that every ideal of a certain set $P(R)$ of a semigroup R is zero, then, in particular, $P(A) = 0$ implies that $P(T(\text{ideals of } B)T) = 0$. Since $TU = U$, then $HT = 0$ for a right or left ideal H of B if and only if $H = 0$. Thus, $THT = 0$ if and only if $TH = 0$. The other parts are consequences.

If B is semiprime, then $TI = 0$, for some right ideal I of B implies that $(IT)^2 = 0$, and $IT = 0$, so $I = 0$. Thus, in this case, B is prime if and only if A is prime. Furthermore, then A is simple if and only if T is the least ideal of B. In this case, if $I = I^2$ is a right ideal contained in T, then

$$I \supseteq IT \supseteq I^2 = I.$$

So $I = IT$.

Conversely, $I = IT$ implies

$$I^2 = I(TIT) = IT = I$$

since TIT must equal T when $I \neq 0$. □

1.15 Definition A right B-module M is a **rational extension** of a submodule N provided that given $x, y \in M$, $y \neq 0$, there exists $b \in B$ such that $xb \in N$ and $yb \neq 0$.

$M_A \supseteq N_A$ is an **essential extension** if $\forall m \in M$, $m \neq 0 \Rightarrow mA \cap N \neq 0$.

Remarks (*a*) Any rational extension is an essential extension, and conversely, if the singular submodule is zero. If $M \in \text{mod-}B$, the **singular submodule** of M, $Z(M)$ is defined by $Z(M) = \{m \in M \mid mI = 0 \text{ for some essential right ideal } I \text{ of } B\}$.

(*b*) Let $\bar{B} = \text{Biend}\ \hat{B}_B$, where $\hat{B}_B$ = injective hull of B_B. Then $\bar{B}$ is the **maximal rational extension** of B in the sense that for any rational extension B' of B, there exists an injective ring homomorphism $h: B' \to \bar{B}$ such that $h|_B = 1_B$. $\bar{B}$ is called the **Johnson–Utumi maximal quotient ring** of B. (See Findlay–Lambek [58], Johnson [51], Johnson–Wong [59], or Faith [67*a*], Chap. 8.) We do not require this remark, however.

1.16 Proposition *Let* $A = \text{End}_{B_0} V$, *where* V *is a faithful, projective left* B_0*-module, let* $B = \text{Biend}_{B_0} V$, $U_0 = \text{hom}_{B_0}(V, B_0)$ and $U = \text{hom}_B(V, B)$. *Then,* $U = U_0 B$; *and* ${}_B V$ *is projective. Moreover, if* $T_0 = \text{trace}_{B_0} V$ *and* $T = \text{trace}_B V$, *then* T_0 *is a left ideal of* B *satisfying*

$$T_0{}^2 = T_0 = TT_0 = BT_0$$

and

$$T_0 B = T_0 T = T = T^2.$$

Thus, there is a lattice isomorphism

$$(1) \begin{cases} (\text{lat } B_B)T \to (\text{lat } B_{0_{B_0}})T_0 \\ I \mapsto IT_0 \end{cases}$$

which induces an isomorphism

$$(2)\quad T(\text{lat}\ {}_B B_B)T \to T_0(\text{lat}\ {}_{B_0} B_{0_{B_0}})T_0$$

of multiplicative semigroups.

Finally, every right ideal I *of* B *is a rational extension of* IT_0 *in* mod-B_0. *In particular,* B *is a rational extension of the subring* B_0.

Proof That $U = U_0 B$, and that ${}_B V$ is projective, follow immediately from the dual basis lemma 1.2(2) and 1.7.

Next, $T_0 V = V = TV$, $T_0{}^2 = T_0$ and $T^2 = T$ by 1.3. Thus,

$$T = [V, U] = [T_0 V, U] = T_0[V, U] = T_0 T.$$

Similarly, $T_0 = TT_0$.

Clearly, the map (1) is a lattice isomorphism. Likewise, the induced map (2) is bijective. The identity

$$[T_0(TIT)T_0][T_0(TJT)T_0] = T_0(TIT)(TJT)T_0$$

shows that (2) is an isomorphism of multiplicative semigroups.

Finally, to show that I is a rational extension of IT_0 in mod-B_0, let $a, b \in I$, $b \neq 0$. Then, $bV \neq 0 \Rightarrow bT_0 \neq 0$. Thus, $bt_0 \neq 0$ for some $t_0 \in T_0$ and clearly, $at_0 \in IT_0$. Thus, I is indeed a rational extension as claimed. In particular, B a rational extension of $BT_0 = T_0$ implies B is a rational extension of B_0. □

1.17 DEFINITION An object $M \in$ mod-A is called **finite dimensional**, or is said to satisfy the a.c.c. $\oplus$, if each independent family of submodules of M is finite.

If M is finite dimensional, we write $\dim M < \infty$. It can be shown that when $\dim M < \infty$, the maximal number of independent submodules of M is always the same natural number (cf. Goldie [60]). This number is called the (Goldie) **dimension** of M. Any Noetherian module is finite dimensional.

1.18 PROPOSITION *Let* $A = \operatorname{End}_B V$ *with* ${}_BV$ *finitely generated projective and faithful, and* $U = \hom_B(V, B)$.

Then (1) $\dim V_A = \dim B_B$,

(2) $\dim A_A = \dim U_B$.

Proof (1) Let $V_1 \oplus V_2 \oplus \ldots \oplus V_i$ be a direct sum of A-submodules of V. Then $[V_1, U] + \ldots + [V_i, U]$ is a direct sum of right ideals of B. For, clearly, each $[V_i, U]$ is a right ideal of B. If $\sum b_i = 0$, $b_i \in [V_i, U]$ then $\sum b_i V = 0 \Rightarrow b_i V = 0$, $\forall i \Rightarrow b_i = 0$, $\forall i$ since ${}_BV$ is faithful. Thus, $\dim V_A \leq \dim B_B$. Similarly, $\dim B_B \leq \dim V_A$.

The proof of (2) is analogous. □

2. *Structure of Noetherian simple rings*

We begin this chapter with a characterization (2.6) of when a simple ring A has the structure $A \approx \text{End}_B V$, where V is a balanced, finitely generated projective over an integral domain B. This happens iff A has a right ideal W which is lateral in the sense that the endomorphism ring of W is an integral domain. (See 2.6.) Moreover, in this case, we have $A \approx \text{End}_{B_0} V$, where B_0 is a domain with at most one nontrivial ideal T, and $B_0 = T + C$, where $C = \text{center } B_0 \approx \text{center } A$. The structure theory for a simple ring A with uniform right ideal W is exactly the same only in this case B and B_0 are right Ore domains (2.17). Furthermore, we have the (Goldie) dimension of V^* over B equals that of A over A

$$\dim_B V^* = \dim A_A.$$

Moreover, A is right Noetherian iff B satisfies the a.c.c. on idempotent right ideals contained in T. (See 2.21.)

A simple ring $A \approx \text{End}_B V$ (as above) is similar to a right Ore domain iff some right ideal of B has a simple endomorphism ring (2.20).

Two different representations

$$A \approx \text{End}_{B_i} V_i, \quad i = 1, 2,$$

as endomorphism rings of finitely generated projectives occur only if B_1 and B_2 have isomorphic right quotient fields $Q = Q(B_1) \approx Q(B_2)$, and, identifying Q with $Q(B_2)$, it is true that B_1 and B_2 must be equivalent right orders in Q. Thus, for each isomorphism class $[A]$ of simple rings with uniform ideals, we can associate the class (B) of equivalent orders in a field Q. (See 2.22.) For two orders B_1 and B_2, there exists a similarity (as rings) $B_1 \sim B_2$ only if $B_1 \overset{Q}{\sim} B_2$ (as orders). Furthermore, when B_1 is simple, then equivalence of orders $B_1 \overset{Q}{\sim} B_2$ implies similarity as rings $\forall B_2$ iff B_1 is right hereditary (2.26).

We investigate the structure of simple hereditary rings, and show *interalia* that each right ideal is generated by two elements. (These results hold more generally for hereditary Noetherian prime rings, but we restrict ourselves to the case of simple rings.)

We conclude Chapter 2 by showing that every simple right Goldie ring of right global dimension not exceeding 2 is similar to a right Ore domain. The proof borrows an idea from Bass's characterization of Noetherian rings of global dimension ≤ 2 as those Noetherian rings for which the duals of finitely generated modules are projective.

2.1 DEFINITION Let I be a right ideal of R. The **idealizer** of I, Idealizer $(I) = \{a \in R \mid aI \subseteq I\}$. Clearly, Idealizer (I) is the largest subring of R in which I is an ideal.

2.2 PROPOSITION *If I is a right ideal of R,*

$$\frac{\text{Idealizer }(I)}{I} \approx \text{End}(R/I)_R.$$

Proof The map defined by $a + I \mapsto a_l$ where $a_l(r + I) = ar + I$, $r \in R$ is clearly a surjective ring homomorphism with kernel I. $\square$

Recall that if $X \subseteq R$ and $I = X^\perp = \{r \in R \mid xr = 0, \forall x \in X\}$, I is called a **right annulet** of R or a **right annihilator** of R.

2.3 PROPOSITION AND DEFINITION *A right ideal V of a ring A is* **lateral** *provided that $V \neq 0$ and* End V_A *is a domain. A principal right ideal $V = vA$ is lateral iff*

$$\text{Idealizer }(v^\perp)/v^\perp$$

is a domain.

Proof The canonical isomorphism $vA \approx A/v^\perp$ together with 2.2 suffices. $\square$

2.4 PROPOSITION *In order that a right ideal $V \neq 0$ of a ring A be lateral, any of the following conditions are sufficient:*

(1) *$V = vA$ for some $v \in A$, with $v^3 \neq 0$, such that for any $x \in A$, $x^\perp \supseteq v^\perp \Rightarrow x^\perp = v^\perp$ or $vx = 0$.*

(2) (Koh) *$V = vA$ for some $v \in A$ such that $v^{\perp}$ is a maximal right annulet.*

(3) *$V = vA$, where v is contained in a minimal left annulet of A.*

Proof (1) By 2.3, we must show for a, b in Idealizer $(v^{\perp})$, that $ab \in v^{\perp}$ and $b \notin v^{\perp} \Rightarrow a \in v^{\perp}$. But one verifies:

$$(va)^{\perp} \supseteq v^{\perp} + bA \supset v^{\perp}.$$

Hence $v(va) = v^2a = 0$. Since $v^3 \neq 0$, then $v^{2\perp} = v^{\perp}$, so $a \in v^{\perp}$ as required.

(2) In the proof of (1), we have that $(va)^{\perp} \supset v^{\perp}$, so $(va)^{\perp} = A$, hence $va = 0$, and $a \in v^{\perp}$.

(3) Then $v^{\perp}$ is a maximal right annulet. □

2.5 LEMMA *Let ${}_BV$ be a faithful finitely generated projective module and $A =$ End ${}_BV$. Then V_A is isomorphic to a right ideal of A iff B_B embeds in $V_B{}^*$.* □

2.6 LEAST IDEAL THEOREM† *A ring A is a simple ring with a lateral right ideal iff A is isomorphic to the endomorphism ring of a balanced, finitely generated projective left module V over a domain B such that $T =$ trace ${}_BV$ is the least ideal of B.*

Proof ($\Leftarrow$): By 1.14(6), $A =$ End ${}_BV$ is simple iff $T =$ trace ${}_BV$ is the least ideal of B. Hence, A is simple. Since ${}_BV$ is faithful and projective, we can assume that V_A is a right ideal of A by 2.5. Since ${}_BV$ is balanced, End $V_A = B$ is a domain and hence, V is a lateral right ideal of A.

Conversely, if V is a lateral right ideal, then $B =$ End V_A is a domain. Since V_A is a right ideal of A and A is simple, V_A is balanced and ${}_BV$ is finitely generated and projective, i.e., $A =$ End ${}_BV$ as required. □

By Proposition 1.4, if $A =$ End ${}_BV$, $B =$ End V_A and ${}_BV$ is finitely generated projective and faithful, then $A =$ End ${}_{B_0}V$, where $B_0 = C + T$, $C =$ center $B \approx$ center A, and ${}_{B_0}V$ is

† In the Introduction, we have stated the Least Ideal Theorem as it applies to a simple Goldie ring, and this theorem specializes to that one by taking V to be any uniform right ideal when A is right Goldie. See 2.34(2).

finitely generated projective and faithful. Moreover, $T_0 = \text{trace}_{B_0} V = \text{trace}_B V = T$ and center $B_0 =$ center B. For $T_0 = T$ by the proof of the second part of 1.4, and if $c \in$ center B_0, $b \in B$, $(bc - cb)T = 0$ because $BT = T$ and $ct = tc$, $\forall t \in T$. Hence $bc = cb$, $\forall b \in B$ implying that $c \in$ center B. The opposite inclusion is obvious. Thus, we have,

2.7 SINGLE IDEAL THEOREM† *A simple ring A with a lateral right ideal V is isomorphic to the endomorphism ring of a finitely generated projective left B-module where B is a domain with at most three ideals. In fact, B can be chosen so that $B = C + T$, $T = \text{trace}_B V$ and $C =$ center $B \approx$ center A is a field.* □

Simple rings with uniform right ideals

2.8 DEFINITION A **right order** R of a ring S is a subring of S such that

$$S = \{ab^{-1} \mid a, b \in R, b \text{ a unit of } S\}.$$

2.9 DEFINITION Q is a **right quotient ring** (quoring) **of** R if R is a subring-1 of Q such that:

(Q1) R has regular elements.
(Q2) Every regular element of R is a unit of Q.
(Q3) If $q \in Q$, then $q = xc^{-1}$, with x, regular $c \in R$.

2.10 DEFINITIONS Let R_1 and R_2 be right orders in a quotient ring Q. Then

(1) they are **equivalent**, $R_1 \overset{Q}{\sim} R_2$, if there exist regular elements a_1, b_1, a_2, b_2 of Q such that $a_1R_1b_1 \subseteq R_2$, $a_2R_2b_2 \subseteq R_1$;

(2) they are **right equivalent**, $R \overset{r}{\sim} R_2$, if there exist regular elements a_1, a_2 of Q such that $a_1R_1 \subseteq R_2$, $a_2R_2 \subseteq R_1$;

(3) they are **left equivalent**, $R_1 \overset{l}{\sim} R_2$, if there exist regular elements b_1, b_2 of Q such that $R_1b_1 \subseteq R_2$, $R_2b_2 \subseteq R_1$.

These are equivalence relations. Also, $R_1 \overset{r}{\sim} R_2 \Rightarrow R_1 \overset{Q}{\sim} R_2$ and $R_1 \overset{l}{\sim} R_2 \Rightarrow R_1 \overset{Q}{\sim} R_2$.

† In the Introduction, we have stated the Single Ideal Theorem as it applies to a simple Goldie ring. 2.7 also includes the Same Centers Theorem described in the Introduction.

2.11 LEMMA *Let R be a right order in a quotient ring Q, and let S be a subring-1 of Q.*

(1) *If there exist regular elements a, b, c, d of Q such that $aRb \subseteq S$, $cSd \subseteq R$, then S is a right order in Q, and $R \overset{Q}{\sim} S$.*

(2) *If there exist regular elements a, c of Q such that $aR \subseteq S$, $cS \subseteq R$, then S is a right order in Q, and $R \overset{r}{\sim} S$. Similarly for left equivalence.*

Proof (1) If $q \in Q$, then

$$a^{-1}qa = xc^{-1}$$

for some $x, c \in R$, and then

$$q = (axb)(acb)^{-1},$$

so that S is a right order of Q similar to R.

Proof of (2) is obtained by taking $b = 1$. □

2.12 DEFINITIONS A **maximal $\overset{Q}{\sim}$ right order** is a right order maximal amongst the right orders which are equivalent to it. Maximal $\overset{r}{\sim}$ and maximal $\overset{l}{\sim}$ right orders are defined likewise.

2.13 DEFINITION A ring R is said to be **right Goldie** provided that R satisfies the a.c.c. on right annihilators, the (a.c.c.)$^{\perp}$ and dim $R < \infty$.

A right Goldie domain of dimension 1 is called a **right Ore domain**. It is elementary to show that any right Goldie domain has dimension 1.

2.14 THEOREM *The following conditions on a ring R are equivalent:*

(1) *R is semiprime right Goldie;*

(2) *R is semiprime, $Z(R) = 0$ and* dim *$R < \infty$;*

(3) *R has a semisimple right quotient ring S.*

We shall postpone the proof of 2.14 until Chapter 4. There, the orders in simple Artin rings will be discussed at great length.

The following concept plays a fundamental role both in the

study of orders in semisimple rings and the representation theory for simple rings.

2.15 DEFINITION $M \in$ mod-A is called **uniform** if M satisfies any of the following equivalent conditions:

(1) $\forall m, m' \in M, mA \cap m'A \neq 0$;

(2) $\hat{M}$ = injective hull M is indecomposable;

(3) $\dim M_A = 1$.

2.16 EXAMPLES (1) Simple modules.

(2) If A is any right Ore domain, A is uniform in mod-A.

By Proposition 1.18(1), if $A = \text{End}\,{}_BV$ where ${}_BV$ is finitely generated projective and faithful, $\dim B_B = \dim V_A$. Thus, B is right Ore iff V_A is uniform. In particular, if A is simple and U_A is a right ideal of A, $K = \text{End}\,U_A$ is right Ore iff U_A is uniform. Moreover, whenever U_A is uniform, any right ideal J of A with endomorphism ring a domain is contained in U_A. For, if $K' = \text{End}\,J_A$ is a domain, since $A = \text{End}\,{}_{K'}J$, $\dim K'_{K'} = \dim J_A \leq \dim A_A < \infty$. Thus, J_A is uniform. Since $VJ \neq 0$, $vJ \neq 0$ for some $v \in V$. However, JQ is clearly a minimal right ideal of the quotient ring of A. In fact, JQ = injective hull of J_A. Thus $vj \neq 0$, $\forall j \in J$, and hence the map $J \to V$ defined by $j \mapsto vj$ is a monomorphism.

2.17 ENDOMORPHISM RING AND PROJECTIVE MODULE THEOREM† *The following conditions are equivalent:*

(1) *A is a simple ring with uniform right ideal V.*

(2) *A is isomorphic to the endomorphism ring of a finitely generated projective left B_0 module V over a right Ore domain B_0 having center C a field such that either B_0 is simple or $B_0 = C + T_0$, $T_0 = \text{trace}\,{}_{B_0}V$, is a ring possessing precisely* 3 *ideals.*
□

2.18 THEOREM (Hart [67]) *If A is a simple ring, then A has a uniform right ideal iff* $\dim A_A < \infty$. *Moreover, in this case, A*

† 2.17 includes the Endomorphism Ring Theorem and the Projective Module Theorem described in the Introduction. This theorem also contains a statement of the Single Ideal Theorem.

is necessarily a Goldie ring and there exists a right Ore domain K and an integer $n > 0$ such that $A \approx eK_ne$ where e is an idempotent belonging to K_n.

Proof $\Leftarrow$: Clear.

$\Rightarrow$: By the preceding Proposition, $A \approx \text{End}\,{}_BV$, ${}_BV$ finitely generated projective and B right Ore. By Proposition 1.18(2), $\dim A_A = \dim V_B{}^* < \infty$.

If U is a uniform right ideal of A, then U_A is a generator and hence, $U^n \approx A \oplus B$ for some integer $n > 0$. Let $e \in \text{End}\, U^n$ be the projection idempotent $e: U^n \to A \to 0$. If $K = \text{End}\, U_A$ then $A \approx eK_ne$. $\square$

2.19 Lemma *Let $A = \text{End}\,{}_BV$, $B = \text{End}\, V_A$ be right Ore with quotient field D, and let ${}_BV$ be finitely generated projective and faithful. If $W_A \subseteq V_A$ and $I = [W, U]$ where $U =$* $\hom_B(V, B)$, $(I : I) = \{d \in D \mid dI \subseteq I\} \approx \text{End}\, I_B$ *and* $\text{End}\, I_B \overset{g}{\approx}$ $\text{End}\, W_A$ *where g is defined by*

$$g(f)(xv) = f(x) \cdot v \quad \forall f \in \text{End}\, I_B, \quad x \in I, v \in V.$$

Proof Clearly, each element $d \in (I : I)$ induces, by left multiplication, an element d_l of $\text{End}\, I_B$, and hence, a ring homomorphism of $(I : I)$ into $\text{End}\, I_B$. To exhibit its inverse, $f \in \text{End}\, I_B$ extends to an $\hat{f}: \hat{I} \to \hat{I}$, and $\hat{f}(1) = d_f$ satisfies $d_f i = \hat{f}(i) = f(i)$, $\forall\ i \in I$, so $d_f \in (I : I)$. The desired inverse is defined by $f \mapsto d_f$. The proof of the second isomorphism is similar. $\square$

2.20 Simple Endomorphism Rings Theorem *Let A be a simple ring with a uniform right ideal V and let D denote the right quotient field of the domain $B = \text{End}\, V_A$. The conditions listed below are equivalent:*

(1) *A is isomorphic to the endomorphism ring of a finitely generated projective left R-module U where R is a simple right Ore domain.*

(2) *A is isomorphic to the endomorphism ring of a finitely generated projective left S-module U, with S a right Ore domain such that the endomorphism ring of some right ideal of S is a simple ring.*

(3) *Some right ideal of B has simple endomorphism ring.*

(4) *The endomorphism ring of some A-submodule of V is simple.*

(5) *The endomorphism ring of some uniform right ideal of A is simple.*

(6) *There is a right ideal I of B and a simple subring G of D such that $GI \subseteq I$ and D is the right quotient field of G.*

(7) *There is a progenerator P in* mod-A *such that* $L = \text{End } P_A$ *is a simple domain, that is, $A \sim L$.*

Proof (1) $\Rightarrow$ (2): Trivial.

(2) $\Rightarrow$ (5): $A = \text{End }_S U$. Therefore, by Lemma 2.5 U_A is isomorphic to a uniform right ideal of A. By the remarks preceding 2.17, $U_A \subseteq V_A$. If I is a right ideal of S such that End I_S is simple, then IU_A is a uniform right ideal of A and End $I_S \approx$ End IU_A by Proposition 2.19.

(5) $\Rightarrow$ (4): Trivial.

(4) $\Rightarrow$ (3): Suppose $Y_A \subseteq V_A$ has a simple endomorphism ring. There exists a right ideal I of B such that $IV = Y$ ($I = [Y, V^*]$) and by Proposition 2.19, End $Y_A \approx$ End I_B.

(3) $\Rightarrow$ (6): If End I_B is simple, $G = (I : I) = \{d \in D \mid dI \subseteq I\} \approx$ End I_B and D is clearly the right quotient field of G.

(6) $\Rightarrow$ (7): Suppose $G \subseteq (I : I) =$ End I_B. Clearly, $P = IV_A$ is uniform and

$$\text{End } I_B \approx \text{End } P_A.$$

If J is any nonzero ideal of $(I : I)$, $J \cap G \neq 0$ since D is an essential extension of G. Therefore, $1 \in J$ which implies that $J = (I : I)$, that is $(I : I)$ is simple. Since A is simple, P_A is a generator in mod-A and therefore, finitely generated projective over $L =$ End P_A. Simplicity of L implies that P is an (L, A) progenerator.

(7) $\Rightarrow$ (1): Trivial. □

2.21 Idempotent Right Ideal Theorem *A ring A is a simple right Noetherian ring iff A is isomorphic to* End ${}_{B_0}V$, *${}_{B_0}V$ finitely generated projective, B_0 a right Ore domain having center C a field and such that either B_0 is simple or $B_0 = C + T_0$, $T_0 =$* trace ${}_{B_0}V$, *is a ring possessing precisely 3 ideals and satisfying the a.c.c. on idempotent right ideals contained in T_0. Moreover, when this is so, $B =$* End V_A *satisfies the a.c.c. on principal right ideals.*

Proof $\Rightarrow$: Clearly, A contains a uniform right ideal V and hence the first part of the necessity follows from 2.17. By 1.14(6) and 1.13(1), B_0 satisfies the a.c.c. on idempotent right ideals contained in T_0.

$\Leftarrow$: First, A is simple by 1.14(6). Next, the a.c.c. on idempotent right ideals contained in T_0 implies that V_A is a Noetherian module by 1.14(6) and 1.13(1): hence, so is A_A since $V^n \approx A \oplus B$ for some $n > 0$.

If $x_1 B \subseteq x_2 B \subseteq \ldots$ is an ascending chain of principal right ideals of B, then

$$x_1 V \subseteq x_2 V \subseteq \ldots \subseteq x_n V \subset \ldots$$

is an ascending chain in V_A which must terminate since V_A is Noetherian. Thus, for some $k > 0$

$$x_k V = x_{k+j} V \quad \forall j \geq 0.$$

Now B is a right Ore domain having right quotient field D say. Thus, $d = x_{k+1}{}^{-1} x_k \in D$ which satisfies $dV = V$ and $dT = T$ since $[x_k V, U] = x_k T = [x_{k+1} V, U] = x_{k+1} T$ and $TU = U$. Since d induces an epimorphism of the Noetherian module V_A, d is a unit of $B \Rightarrow x_k B = x_{k+1} B$. $\square$

Two questions naturally arise pertaining to the above representation of A:

(1) Is the right Ore domain B necessarily Noetherian?

(2) If $A \approx \mathrm{End}_{B'} V'$, where $_{B'}V'$ is finitely generated projective and B' right Ore, what is the relationship between B and B'?

The answer to (1) is 'no' as will be shown in Chapter 4. The next Proposition gives us some indication of the relationship between B and B'.

2.22 Uniqueness of Orders Theorem *Let B_i be a right Ore domain with right quotient field D_i, and U_i finitely generated projective in B_i-mod, and $A_i = \mathrm{End}_{B_i} U_i$, $i = 1, 2$. If there is a ring isomorphism $f: A_1 \to A_2$, then there is an isomorphism $g: D_1 \to D_2$ such that $g(B_1)$ and B_2 are equivalent right orders in D_2.*

Proof By 2.5, U_i is canonically isomorphic to a uniform right ideal of A_i. Therefore, $f(U_1) = V_2$ is a uniform right ideal of A_2. In fact, V_2 is isomorphic to an A_2-submodule W_2 of U_2. Moreover, $W_2 = IU_2$ where $I = [W_2, U_2{}^*]$ is a right ideal of B_2 and End $W_{2_{A_2}} \approx$ End $I_{B_2} \approx \{d \in D_2 \mid dI \subseteq I\} = B$. Clearly, B is a right order of D_2 and in fact, equivalent to B_2. For, if $x \in I$, $x \neq 0$, then $Bx \subseteq B_2$ and $xB_2 \subseteq B$.

Now, $R =$ Biend ${}_{B_2}W_2 =$ End $W_{2_{A_2}}$ is a right Ore domain containing B_2 by 1.16 and since $RT \subseteq B_2$ and $B_2 \subseteq R$, where $T =$ trace ${}_{B_2}W_2$, R and B_2 have the same right quotient field D_2 and are equivalent right orders. Similarly, $S =$ Biend $U_{1_{A_1}}$ and B_1 are equivalent right orders of D_1. Let $g : D_1 \to D_2$ be the unique extension of the isomorphism $S \to B$ (defined by f). Since S is equivalent to B_1 and B is equivalent to B_2, $g(S) = B$ is equivalent to $g(B_1)$ and hence, $g(B_1)$ is equivalent to B_2. □

Having established necessary and sufficient conditions for a simple ring A to be similar to a right Ore domain B, it is instructive to analyze the structure of the ring A for low global dimensions.

Zero global dimension

By definition, the right global dimension of a ring A is zero iff every right A-module is projective (injective), and this is equivalent to stating that every right A-module is semisimple in the sense that every module is a direct sum of simple modules. By the Wedderburn–Artin Theorem, these rings are finite products of matrix rings over fields. (See, for example Faith [73*a*, pp. 369–70, 8.8 and 8.11].) We next derive the Wedderburn–Artin theorem for simple rings using our methods.

2.23 Theorem (Wedderburn–Artin) *A ring A is simple Artinian iff for some $n > 0$, $A \approx B_n$, B a field.*

Proof $\Leftarrow$: Clear.

$\Rightarrow$: By 2.21, $A \approx$ End ${}_BV$, ${}_BV$ finitely generated projective, B a right Ore domain satisfying (2) of 2.17 and the a.c.c. on idempotent right ideals contained in $T =$ trace ${}_BV$.

By 1.13(1) and 1.14(6), V_A is a Noetherian module. Thus, for some $n > 0$, there exists an epimorphism $A_A{}^n \to V_A \to 0$ which implies that V_A is an Artinian module $\Rightarrow B$ satisfies the descending chain condition (d.c.c.) on idempotent right ideals contained in $T \Rightarrow \text{Soc } B_B \neq 0 \Rightarrow B$ is a field, i.e., $\text{Soc } A_A \neq 0 \Leftrightarrow \text{Soc } B_B \neq 0$. Thus, B is a field, ${}_BV$ is a finite dimensional vector space over B, hence ${}_BV \approx B^n$ for some $n > 0 \Rightarrow A \approx B_n$. □

By the symmetrical properties of matrix rings one sees that A has zero right global dimension iff A has zero left global dimension, since A is right semisimple iff A is left semisimple.

Right global dimension of $A = 1$

A ring A is called **right hereditary** if the right global dimension of A equals 1; equivalently, each right ideal of A is projective.

2.24 LEMMA *Any finite dimensional right hereditary ring R is right Noetherian.*

Proof Let I be any right ideal of R. Clearly, by finite dimensionality of R, I contains a finitely generated essential right ideal J. Since I is projective, I is a summand of $R^{(X)}$ for some X, and by finite generation of J, J is contained in a finite free summand of $R^{(X)}$, say $R^{(n)}$. Since $\text{hom}_R(I/J, R) = 0$, any homomorphism $f: I \to R$ satisfying $f|J = 0$ must already be 0. Thus, $I \subseteq R^{(n)}$ and is a summand. □

Thus, a simple right hereditary ring with a uniform right ideal is similar to a right Ore domain which is necessarily Noetherian. (For r.gl.dim $= 2$, see 2.40.) The following proposition provides a characterization of this class of rings in terms of the uniform right ideals.

2.25 HEREDITARY RING THEOREM *Let A be a simple ring with a uniform right ideal U. Then the following conditions are equivalent:*

(1) *The endomorphism ring of every nonzero uniform right ideal V of A is simple.*

(2) *Every uniform right ideal V of A is finitely generated projective.*

(3) *A is right hereditary and right Noetherian.*

Proof (1) $\Leftrightarrow$ (2): Let $B = \text{End}\, V_A$. Since V_A is a generator, ${}_BV$ is finitely generated projective and a generator since B is simple. Thus, V_A is finitely generated projective since $A = \text{End}\, {}_BV$. The converse is similar.

(1) $\Rightarrow$ (3): Let U be any uniform right ideal of A. Then U_A is finitely generated projective by (1) $\Rightarrow$ (2). Thus, every A-submodule of U is finitely generated projective implying that U_A is Noetherian. Since U_A is a generator, $U^n \approx A \oplus X$ for some $X \in \text{mod-}A$ implying that A_A is likewise Noetherian.

By an easy induction on l, every A-submodule of U^l is finitely generated projective. Thus, every right ideal of A is (finitely generated) projective or A is right hereditary.

(3) $\Rightarrow$ (1): If A is right hereditary and Noetherian, and W is any uniform right ideal of A, W_A is an A-progenerator implying that $A \sim B = \text{End}\, W_A$. Thus, B is simple. □

In general we have observed that there is no precise relationship between an order A and an equivalent order B. However, when A is a simple, Noetherian, hereditary domain, there is a far more fundamental connection. Namely,

2.26 PROPOSITION *Let B be a right Ore domain with right quotient field D, and let A be a subring of D which is a right order in D. Consider the conditions:*

(1) *$A \sim B$.*

(2) *There is a right ideal I of B and a ring isomorphism $A \approx (I : I) = \{d \in D \mid dI \subseteq I\}$.*

(3) *A and B are equivalent right orders.*

Then (1) $\Rightarrow$ (2) $\Rightarrow$ (3).

If B is a simple ring, then every ring satisfying (2) *or* (3) *is simple only if B is right hereditary and right Noetherian. In this case,* (1) $\Leftrightarrow$ (3) *for any maximal order A.*

Proof In order to establish (1) $\Rightarrow$ (2) $\Rightarrow$ (3), we first prove the following lemma:

2.27 LEMMA *Let B_i, $i = 1, 2$, be right Ore domains having the same right quotient field D. If $B_1 \sim B_2$, then B_1 and B_2 are* *equivalent right orders and there is a right ideal I of B_2 together with a ring isomorphism*

$$B_1 \approx (I : I).$$

Proof By 1.10, $B_1 = \text{End } P_{B_2}$ and $B_2 = \text{End }_{B_1}P$ for some (B_1, B_2)-bimodule progenerator P. By 2.5 P is isomorphic to a right ideal I of B_2 and clearly, $B_1 \approx (I : I)$. For any $x \in I$, $B_1 x \subseteq I \subseteq B_2$ and $xB_2 \subseteq I \subseteq B_1$. Thus, B_1 and B_2 are equivalent right orders. □

Now assume that B is simple. If I is any nonzero right ideal of B, I is a left ideal of the ring $A = (I : I)$ and A is a right order of D equivalent to B. $A \sim B$ implies that A is simple. Thus, I is a generator in A-mod and mod-B and by 1.7 finitely generated projective in both. Thus, B is right hereditary and right Noetherian provided that every equivalent right order of D is simple.

Conversely, assuming that B is a simple, right hereditary, right Noetherian domain, let A be a maximal right order equivalent to B. Then, $xAy \subseteq B$ for nonzero elements $x, y \in B$. Thus, $xy = b \in B$ and $y = x^{-1}b$. Set $A' = xAx^{-1}$. Clearly, $A'b \subseteq B$ and hence $A'I \subseteq I$ for $I = A'bB$. Let $S = (I : I) \approx \text{End } I_B$. Since I_B is a progenerator, $S \sim B$ and hence, is simple. Since A is a maximal order, $A = x^{-1}Sx$ and thus, A is likewise simple. Since $A \sim S$, $A \sim B$. □

In a different direction, we have the following result which has as an immediate corollary the fact that each right ideal of A in 2.25(3) can be generated by 2 elements. Thus emerges an intimate connection between these rings and the classical Dedekind domains.

2.28 THEOREM (Webber [70]) *Let A be a simple, right hereditary, right Noetherian ring. Let I, J and K be any three nonzero right ideals of A, J essential in A. Then, there exists a right ideal L such that $I \oplus K \approx J \oplus L$.*

Proof If $I \approx J$, we are finished. Otherwise, choose $j_0 \in J$ which is regular and set $J_0 = j_0 I$. Since $I \approx J_0$, if we set $L_0 = K$,

$$I \oplus K \approx J_0 \oplus L_0.$$

Moreover, $J_0 \subset J$. Thus, $\{a \in A \mid Ja \subseteq J_0\}$ must be zero since this set is a 2-sided ideal of A, and A is simple while $JA = J \supset J_0$. Hence, $\exists j_1 \in J$ such that $j_1 L_0 \not\subset J_0$. Set $J_1 = J_0 + j_1 L_0$. We have an epimorphism

$$\phi : L_0 \to j_1 L_0 / J_0 \cap j_1 L_0 = J_1 / J_0.$$

Let $L_1 = \ker \phi$. Then $L_1 \oplus J_1 \approx J_0 \oplus L_0$ by Schanuel's Lemma. The ascending chain $J_0 \subset J_1 \subset \ldots$ must terminate in finitely many steps. □

2.29 COROLLARY *Each right ideal of a simple right hereditary right Noetherian ring A can be generated by 2 elements.*

A duality for torsion modules

Let A be a right Goldie ring and $M \in$ mod-A. An element $x \in M$ is called a **torsion element** in case $xc = 0$ for some regular element $c \in A$. Using the fact that A is a right Goldie ring, the set $t(M)$ of torsion elements of M can be seen to be a submodule of M, called its **torsion submodule**. The proof of this requires the theorem of Goldie and Lesieur–Croisot (given in Chapter 4) which states that A has a right quotient ring $Q(A)$, or equivalently, for given elements $a, b \in A$, and b regular, the Ore condition holds:

$\exists a_1 \in A$, regular $b_1 \in A$ such that

$$ab_1 = ba_1.$$

Given that $t(M)$ exists, we say that M is a **torsion module** if $t(M) = M$; M is **torsion-free** if $t(M) = 0$.

2.30 LEMMA *Let A be a semiprime, two-sided hereditary Noetherian ring with quoring Q and I an essential right ideal of A. Then*

(1) $I^* \approx I^{-1} = \{q \in Q \mid qI \subseteq A\}$.
(2) $I^{-1-1-1} = I^{-1}$.
(3) $I^{-1-1} = I$.
(4) *A/I is Artinian.*

Proof The proof of (1) is similar to that of 2.19 while the proof of (2) is trivial. Note that (1) and (2) are valid for any semiprime, two-sided Noetherian ring. Henceforth, we shall freely identify the modules I^* and I^{-1}. Since $I^{**} \approx I$ (proof?) and $I^{-1-1} \subseteq I$, $I^{-1-1} = I$.

To show that A/I is Artinian, observe that I contains a regular element c since it is essential. (See 4.4 for a proof of this fact.) If

$$A \supseteq I_1 \supseteq I_2 \supseteq \ldots \supseteq I \supseteq cA$$

is a descending chain containing I,

$$Ac^{-1} = (cA)^{-1} \supseteq I^{-1} \supseteq \ldots \supseteq I_n{}^{-1} \supseteq \ldots \supseteq I_1{}^{-1} \supseteq A$$

is an ascending chain contained in Ac^{-1}. Since A is left Noetherian, for some n, $I_n{}^{-1} = I_{n+j}{}^{-1}$, $\forall j \geq 1$. Thus, $I_n = I_{n+j}$, $\forall j \geq 1$ by (2). □

2.31 DEFINITION Let $\mathscr{C}$ and $\mathscr{C}'$ be two categories, $T : \mathscr{C} \rightsquigarrow \mathscr{C}'$ and $S : \mathscr{C}' \rightsquigarrow \mathscr{C}$ be two contravariant functors satisfying $T \circ S \simeq 1_{\mathscr{C}'}$, and $S \circ T \simeq 1_{\mathscr{C}}$. Then the pair (T, S) is called a **duality** between $\mathscr{C}$ and $\mathscr{C}'$ and the categories $\mathscr{C}$ and $\mathscr{C}'$ are said to be **dual** to one another.

2.32 PROPOSITION *Let A be a semiprime, two-sided hereditary Noetherian ring and let $\mathscr{T}_A$* (resp. ${}_A\mathscr{T}$) *denote the full subcategory of* mod-A *(resp.* A-mod) *consisting of all finitely generated torsion modules. Then, the functor* $\hom_A(-, K)$ *defines a duality between $\mathscr{T}_A$ and ${}_A\mathscr{T}$ where $K = Q/A$, and Q is the two-sided quotient ring of A.*

Proof By 2.30, each cyclic and hence, each finitely generated torsion module is Artinian and thus, has finite length. If A/M is a simple right or left A-module, then it is trivial to check that

$$(A/M)^{\#} = \hom_A(A/M, K) \approx M^{-1}/A$$

and is simple. Moreover $(M^{-1}/A)^{\#} \approx A/M$. Now, if M has finite length n, let N be a submodule of length $n - 1$. If we apply $^{\#}$ twice to the exact sequence

$$0 \to N \to M \to M/N \to 0$$

we obtain the following commutative diagram by injectivity of K (on both sides):

$$\begin{array}{ccccccccc}
0 & \longrightarrow & N & \longrightarrow & M & \longrightarrow & M/N & \longrightarrow & 0 \\
 & & \downarrow & & \downarrow & & \downarrow & & \\
0 & \longrightarrow & N^{\#\#} & \to & M^{\#\#} & \to & M/N^{\#\#} & \to & 0
\end{array}$$

By induction on n, $N^{\#\#} \approx N$ and ditto for M/N since this module is necessarily simple. Thus, $M^{\#\#} \approx M$ completing the proof. □

2.33 COROLLARY *Under the hypothesis of Proposition* 2.32, *each object in* $\mathcal{T}_A$ *(resp.* ${}_A\mathcal{T}$*) has a composition series of finite length.* □

Structure of projectives over simple hereditary rings

Let M be an arbitrary finitely generated right A-module. Our goal is to show that $M \approx M_1 \oplus M_2$ where M_1 is projective (in fact $M_1 \approx A^n \oplus I$ for some $n > 0$ and right ideal I), and M_2 is the torsion submodule of M. Since we know the structure of M_2, we shall have gained a good handle on the subcategory of all finitely generated A-modules.

Recall that if I is any right ideal of A, there exists a right ideal K of A maximal with respect to the property that $I \cap K = 0$ (Zorn). Such a K is called a **relative complement** of I. The set of all K which are relative complements of some right ideal I of A are called the **complement right ideals** of A.

2.34 PROPOSITION (Goldie [60]) *Let A be a semiprime right Goldie ring. Then*

(1) *each right annulet of A is a complement right ideal;*
(2) *if U is uniform and $0 \neq u \in U$, $au = 0 \Rightarrow aU = 0$;*
(3) *if I is any right annulet, $I = a^{\perp}$ for some $a \in A$;*
(4) *each minimal right annulet is uniform.*

Proof (Goldie) (1) Let I be any right annulet, K a relative complement and $I' \supseteq I$ a relative complement of K. Since $I \oplus K$ is essential,

$$(x : I \oplus K) = \{a \in A \mid xa \in I \oplus K\}$$

is essential in A for all $x \in A$. If $x \in I'$, $x(x : I \oplus K) \subseteq I' \cap I \oplus K = I$ and hence ${}^{\perp}Ix(x : I \oplus K) = 0$. Since the singular ideal of A, $Z(A) = 0$, then ${}^{\perp}Ix = 0 \Rightarrow x \in I$.

(2) We must show that ${}^{\perp}U = \{x \in A \mid x^{\perp} \cap U \neq 0\}$. Set $X = $ R.H.S. Claim: $XU \subseteq Z(A)$. Since $Z(A) = 0$, this will imply that $X = {}^{\perp}U$.

Let $x \in X$, $u \in U$ and I a right ideal of A. If $uI = 0$, then $xuI = 0 \Rightarrow (xu)^{\perp} \cap I \neq 0$. If $uI \neq 0$, then $x^{\perp} \cap U \cap uI \neq 0$ by uniformity of $U \Rightarrow (xu)^{\perp} \cap I \neq 0$. Thus, $(xu)^{\perp}$ is essential $\Rightarrow xu \in Z(A)$.

(3) Let I be any right annulet and U a uniform right ideal satisfying $I \cap U = 0$. We claim there exists $u \in U$ such that $u^{\perp} \supseteq I$ and $u^{\perp} \cap U = 0$. For, consider ${}^{\perp}I \cap U$. If ${}^{\perp}I \cap U = 0$ $\Rightarrow U^{\perp}I \subseteq {}^{\perp}I \cap U = 0 \Rightarrow ({}^{\perp}IU)^2 = 0 \Rightarrow {}^{\perp}IU = 0$ by semiprimeness. Hence, $U \subseteq ({}^{\perp}I)^{\perp} = I$, a contradiction. Thus, we can choose $u \in {}^{\perp}I \cap U$, $u \neq 0$ such that $u^{\perp} \supseteq I$ (obvious) and $u^{\perp} \cap U = 0$ by a similar argument.

Let I be any right annulet and K a relative complement. Choose $U_1 \subseteq K$ uniform. Since $I \cap U_1 = 0$ there exists $u_1 \in U_1$ such that $u_1{}^{\perp} \supseteq I$ and $u_1{}^{\perp} \cap U_1 = 0$. If $u_1{}^{\perp} \cap K = 0$ we stop. Otherwise, choose $U_2 \subseteq u_1{}^{\perp} \cap K$ uniform, $u_2 \in U_2$ satisfying $u_2{}^{\perp} \supseteq I$ and $u_2{}^{\perp} \cap U_2 = 0$. If $u_1{}^{\perp} \cap u_2{}^{\perp} \cap K = 0$ we stop, otherwise continue. Eventually, since the U_i are independent and $\dim A < \infty$, there exists an integer n such that $u_1{}^{\perp} \cap \ldots \cap u_n{}^{\perp} \cap K = 0$. Then by (1), $I = u_1{}^{\perp} \cap \ldots \cap u_n{}^{\perp}$ and clearly, $I = a^{\perp}$ for $a = u_1 + \ldots + u_n$.

(4) Clear. □

2.35 Lemma *Let A be a simple, right hereditary, right Noetherian ring. Then A is isomorphic to a direct sum of uniform right ideals of A. More generally, each projective right A-module is isomorphic to a direct sum of uniform right ideals of A.*

Proof Recall that if $A \approx \sum_{j=1}^{n} \oplus I_j$, I_j, $j = 1, \ldots, n$, a right ideal of A, and P_A is projective, $P \approx \sum_{i \in J} \oplus J_i$ where each J_i is a right ideal of A contained in some I_j (see Cartan–Eilenberg [56], p. 13, Theorem 5.3 or Lambek [66], p. 85). By 2.34, there exists a uniform right ideal $U = a^{\perp}$ for some $a \in A$. Thus, by projectivity of aA,

$$A \approx U \oplus aA.$$

By induction, $A \approx \sum_{k=1}^{m} \oplus U_k$ where each U_k is a uniform right ideal of A. Thus, each projective right A-module can be similarly decomposed. □

In 2.36 and 2.37 which follow, A will denote a two-sided hereditary, Noetherian simple ring.

2.36 PROPOSITION *Let M be an arbitrary finitely generated torsion free A-module. Then M is projective. In fact, $M \approx A^n \oplus I$ for some right ideal I of A. Moreover, any nonfinitely generated projective A-module is free.*

Proof Clearly, $M \hookrightarrow M \oplus {}_A Q$ which is finitely generated and torsion free over Q. Thus, $M \otimes {}_A Q \hookrightarrow Q^n$ and hence, elements of M can be thought of as n-tuples of elements of Q. Multiplying by a common (left) denominator embeds M in A^m for some $m > 0$.

By 2.35 we can assume M is isomorphic to a direct sum of uniform right ideals of A. If dim $A = n$, any direct sum of $m \leq n$ uniform right ideals is clearly isomorphic to a right ideal of A which is essential iff $m = n$. Hence, $M \approx \sum_{j=1}^{p} \oplus I_j \oplus I'$ where each I_j is essential and I' is some right ideal of A. By repeated applications of 2.28, $M \approx A^p \oplus I$, I a right ideal of A. The final statement, which we leave for the reader, is an easy application of 2.28. □

2.37 THEOREM *Let $M \in$ mod-A be finitely generated. Then $M \approx t(M) \oplus M'$ where M' is torsion free.* □

All of the results presented thus far in this section hold for a more general class of rings, namely, the Dedekind prime rings. The reader should consult Eisenbud–Robson [70*a*,*b*] for a thorough exposition of the properties of this important class of rings.

Rings of global dimension 2

In this section we prove that simple right Goldie rings of right global dimension ≤ 2 are all similar to right Ore domains. Hence, in this case at least the underlying domains are also simple (and right Noetherian! cf. 2.25).

2.38 PROPOSITION *If U is any right annulet of A where* r.gl.dim $A \leq 2$, *then U is projective.*

Proof Since $U = a^{\perp}$, $A/U \hookrightarrow A$ via the map f defined by

$$1 + U \mapsto a.$$

Consider the following diagram:

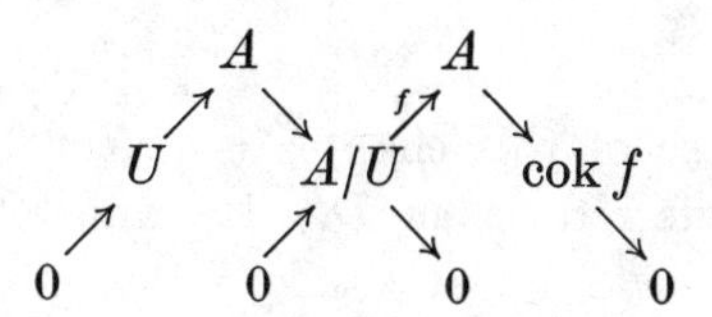

Since r.gl.dim $A \leq 2$, U_A is projective. □

2.39 LEMMA *Any projective uniform right ideal U of A is finitely generated.*

Proof Just parrot the proof of 2.24 using the obvious fact that any cyclic submodule of U is an essential submodule. □

2.40 GLOBAL DIMENSION 2 THEOREM *Let A be a simple right Goldie ring of* r.gl.dim $A \leq 2$. *Then A is similar to a right Ore domain.*

Proof Let U_A be a minimal right annulet of A. By 2.38 and 2.39, U_A is finitely generated projective and necessarily, a generator by simplicity of A. Thus, $A \sim B = \text{End } U_A$ by 1.12. □

3. *Noetherian simple domains*

The main purpose of this chapter is to construct the simple Ore domains, referred to as D_1, D_2, and D_3 in the Introduction.

Without going into great detail, these rings are: (D_1) the ring $\mathscr{D}_R$ of differential polynomials in commuting derivations $\delta_1, \ldots, \delta_n$ over a simple Ore domain R; (D_2) the skew polynomial ring $A = k[x, \rho, \delta]$ with respect to a ρ-derivation δ (and suitable localizations of A); and (D_3), the ring $D \otimes {}_C C(x)$, where D is a transcendental central division algebra over C, and $C(x)$ is the field of rational functions over C in a variable x.

These provide us with the models needed in order to venture further in the structure theory of simple rings in Chapter 4 and the following chapters.

Differential polynomial rings

A **derivation** of a ring R is an additive homomorphism $\delta : R \to R$ such that

$$\delta(ab) = \delta(a)b + a\delta(b)$$

$\forall a, b \in R$. For noncommutative rings, we have, for each x not in the center of R, an **inner derivation** D_x defined by

$$D_x(a) = ax - xa \quad \forall a \in R.$$

Also, any derivation δ of R extends to a derivation Δ of the (ordinary) polynomial ring $R[x]$ by setting $\Delta(x) = 1$. Thus,

$$\Delta\left(\sum_{i=0}^{n} a_i x^i\right) = \sum_{i=1}^{n} i\delta(a_i)x^{i-1}.$$

Let δ be a derivation of R, and let $\mathscr{D}_R$ be the set of all polynomials in the symbol δ. Then $\mathscr{D} = \mathscr{D}_R$ is a ring with respect to the operations of addition of polynomials, and multiplication induced by the relation

$$\delta a = a\delta + \delta(a) \quad \forall a \in R.$$

$\mathscr{D}$ is called the **ring of differential polynomials in δ over R** or the **ring of linear differential operators** (l.d.o.) **with coefficients in R.** If $a = \sum_{i=0}^{n} a_i \delta^i \in \mathscr{D}$, $a_n \neq 0$, we define the **degree** of a, deg a, to be n if $n \geq 0$; $-\infty$ otherwise. If R is a domain, then one readily obtains the formula

$$\deg(ab) = \deg a + \deg b \quad \forall a, b \in \mathscr{D}.$$

3.1 PROPOSITION *If R is left (right) Noetherian, so is $\mathscr{D}$.*

Proof The argument used in the proof of the classical Hilbert Basis Theorem is applicable, *mutatis mutandis.* □

When R is a field, even more can be said: let $a, b \in \mathscr{D}$. Then there exist $q, r \in \mathscr{D}$, $0 \leq \deg r < \deg b$, such that

$$a = qb + r.$$

Similarly, there exist $q', r' \in \mathscr{D}$, $0 \leq \deg r' < \deg b$, such that

$$a = bq' + r'.$$

To demonstrate the above, we can clearly assume that $\deg a \geq \deg b$. Suppose that the leading coefficient of a equals a_n and that of b equals b_m. Clearly,

$$\deg(a - a_n \delta^{n-m} b_m{}^{-1} b) < n.$$

The proof is now completed by induction. As we shall see later, $\mathscr{D}$ need not be a pli-domain (principal left ideal domain) whenever R is.

The importance of $\mathscr{D}$ in the study of simple rings begins to emerge with the next proposition.

3.2 THEOREM *Let R be an ordinary differential ring which is also a $\mathbb{Q}$-algebra.* (*R is called a* **Ritt algebra**.) *Then the following are equivalent:*

(*a*) *$\mathscr{D}$ is simple.*

(*b*) *δ is outer and R has no δ-invariant ideals other than* 0 *and R. An ideal I of R is called* **δ-invariant** *if $\delta(I) \subseteq I$.*

Proof (b) $\Rightarrow$ (a): Suppose $0 \subset I \subset \mathscr{D}$ is a non-zero ideal of $\mathscr{D}$. Let $t = \sum_{i=0}^{n} r_i\delta^i$ be an element of I of least non-negative degree. We shall show that $n = 0$. Let

$$t = r_n\delta^n + p_t(\delta),$$

where $\deg p_t(\delta) < n$. The set L of all $r \in R$ such that either $r = 0$ or r is the leading coefficient of an element in I of degree n is clearly a δ-invariant ideal of R. Thus, $L = R$, which implies that there exists an element of the form $t = \delta^n + p_t(\delta)$ in I. (Note that the leading coefficient of t equals 1.)

Next, for all $r \in R$, $tr - rt \in I$ and has degree less than n. Therefore, $tr = rt$, $\forall r \in R$, implying that t belongs to the centralizer of R in $\mathscr{D}$. Moreover, if

$$t = \delta^n + r_{n-1}\delta^{n-1} + \ldots + r_0, tr = rt \quad \forall r \in R \tag{1}$$

implies that

$$n\delta(r) + r_{n-1}r = rr_{n-1} \quad \forall r \in R$$

by equating the coefficients of both sides of (1). Thus, if $n \neq 0$, δ is inner, contradicting (*b*). Thus, $n = 0$, implying that $I = \mathscr{D}$.

(a) $\Rightarrow$ (b) Suppose $\mathscr{D}$ is simple and δ is defined to be the inner derivation determined by a, D_a, i.e., $\delta(r) = ar - ra$, $\forall r \in R$.

Then it is trivial to show that $\mathscr{D}(\delta + a)$ is a two-sided ideal of $\mathscr{D}$ and clearly, $\neq \mathscr{D}$.

If I is a nontrivial δ-invariant ideal of R, then the set

$$\bar{I} = \left\{\sum_{i=0}^{n} r_i\delta^i \mid r_i \in I\right\}$$

is clearly a nontrivial ideal of $\mathscr{D}$. Thus, (a) $\Rightarrow$ (b). $\square$

3.2a THEOREM *Suppose R is a commutative domain with the characteristic of R,* char *R, $= p \neq 0$. Then $\mathscr{D}_R$ is simple iff R is a field and $\dim_C R$ is infinite, where $C =$* **constants** *$\delta = \{r \in R \mid \delta(r) = 0\}$.*

Proof $\Rightarrow$: First, $\delta(r^p) = 0$ for all (nonunits) $r \in R$. Assuming that R is not a field, there exists a nonunit $r_0 \in R$ satisfying

$\delta(r_0{}^p) = 0$. Thus, R has a nontrivial δ-invariant ideal, namely $Rr_0{}^pR$, which contradicts the simplicity of $\mathscr{D}_R$. Hence, R must be a field whenever $\mathscr{D}_R$ is simple and char $R = p$. If $\dim_C R$ is finite, then, by Jacobson [64*b*, p. 190, Exercises 3 and 7] $\mathscr{D}_R$ is not simple.

$\Leftarrow$: By the remarks following 3.1 $\mathscr{D} = \mathscr{D}_R$ is a pli (pri)-domain. Let I be a nonzero ideal of $\mathscr{D}$. Then, $I = \mathscr{D}f$ for some $f \in \mathscr{D}$. Suppose $f = \sum_{i=0}^{n} a_i\delta^i$ where we may clearly assume that $a_n = 1$. For each $\alpha \in R$, $f\alpha = \beta f$ for some $\beta \in R$ since I is an ideal. By comparing constant terms, we obtain

$$\sum_{i=0}^{n} a_i\delta^i(\alpha) = \beta a_0.$$

Moreover, by comparing terms of degree n, we obtain $a_n\alpha = \alpha = \beta a_n = \beta$. Now using the easily established fact that $f\delta = \delta f$ and the observation that δ can never be inner, we find that each coefficient of f is either 0 or belongs to the center of R, thus,

$$\sum_{i=0}^{n} a_i\delta^i(\alpha) = a_0\alpha$$

or

$$\sum_{i=1}^{n} a_i\delta^i(\alpha) = 0 \quad \forall\alpha \in R.$$

However, if $n \geq 1$, this readily implies that $\dim_C R$ is finite, a contradiction. $\square$

By parrotting the argument that r.gl.dim $R[X] \leq$ r.gl.dim $R + 1$, we obtain

$$1 \leq \text{r.gl.dim } \mathscr{D} \leq \text{r.gl.dim } R + 1.$$

Thus, when R is semisimple, $\mathscr{D}$ is hereditary. Examples where r.gl.dim $\mathscr{D} <$ r.gl.dim $R + 1$ abound as we shall now show.

An example

Let $R = k[X]$ be the ordinary polynomial ring in one indeterminate, k a field with characteristic equal to 0, and δ be defined as the formal derivative of R. Then, clearly, the hypotheses of

Proposition 3.2 are satisfied. Thus, $\mathscr{D}_R$ is simple. This particular ring, denoted also by A_1, has been the source of many important examples and counterexamples. (See Rinehart [62] and Webber [70].)

Abbreviate $\mathscr{D}_R$ to $\mathscr{D}$. We shall first show that r.gl.dim $\mathscr{D} =$ r.gl.dim R.

3.3 LEMMA *Let P be a nonzero prime ideal of R and I a right ideal of $\mathscr{D}$ satisfying*

$$P\mathscr{D} \subseteq I \subset \mathscr{D}.$$

Then, $I = P\mathscr{D}$. In particular, $\mathscr{D}/P\mathscr{D}$ is simple.

Proof Suppose $I \neq P\mathscr{D}$. Choose $a = \sum_{i=0}^{n} a_i\delta^i \in I$ with $a_i \in R$, $a_n \notin P$ and n minimal with respect to this property. That this choice can actually be made is clear.

Since $a\alpha - \alpha a$ has leading coefficient $n\delta(\alpha)a_n$ $\forall\, \alpha \in P$, by minimality of n, $n\ \delta(\alpha)a_n \in P$ implying that $n\delta(\alpha) \in P$. Since char $R = 0$, we have $\delta(P) \subseteq P$, a contradiction. Hence, $n = 0$ which is again a contradiction. □

3.4 THEOREM *Let I be a right ideal of $\mathscr{D}$. Then there exists a chain of right ideals between I and I^{**} such that each factor is, up to isomorphism, of the form $\mathscr{D}/P\mathscr{D}$, P a prime ideal of R.*

Proof Since $I^{***} = I^*$ (2.30), it suffices to show that if J is a right ideal of $\mathscr{D}$, $J \supset I$, J/I is cyclic, and $J^* = I^*$, a chain of right ideals exists between J and I whose factors are all of the stated form. Since $J^* = I^*$, $I' = IS^{-1} = J' = JS^{-1}$ where S is the set of all nonzero elements of $k[X]$ and $IS^{-1} = \{is^{-1} \mid i \in I, s \in S\}$.

To show this, suppose on the contrary that $I' \subset J'$. Now, $\mathscr{D}S^{-1} = \mathscr{D}_{k(X)}$ and since $\mathscr{D}_{k(X)}$ is a pri-domain and thus, necessarily, $I'^* \supset J'^*$, we can find an element q in the quotient field of $\mathscr{D}$ satisfying $qI' \subset \mathscr{D}S^{-1}$, $qJ' \not\subset \mathscr{D}S^{-1}$. In particular, $qI \subset \mathscr{D}S^{-1}$, $qJ \not\subset \mathscr{D}S^{-1}$. By multiplying q by a suitable element of S, we can further assume that $qI \subset \mathscr{D}$, $qJ \not\subset \mathscr{D}$. However, by 2.30, this implies that $I^* \supset J^*$, a contradiction.

By assumption, $J = a\mathscr{D} + I$ for some $a \in \mathscr{D}$ and since $J' = I'$, it is quite easy to show that

$$\mathrm{Ann}_{\mathscr{D}}(a + I) = \{d \in \mathscr{D} \mid ad \in I\}$$

contains a nonzero ideal T of R. Clearly, J/I is an epimorphic image of $\mathscr{D}/T\mathscr{D}$. Since R/T is Artinian, there exists a chain of ideals T_i between R and T such that T_i/T_{i+1} is simple, in fact, isomorphic to R/P_i where P_i is a maximal ideal of R. Let J_i/I denote the image of $T_i\mathscr{D}/T\mathscr{D}$ under the surjection

$$\mathscr{D}/T\mathscr{D} \to J/I \to 0.$$

$$T_i\mathscr{D}/T_{i+1}\mathscr{D} \approx T_i/T_{i+1} \otimes_R \mathscr{D} \approx R/P_i \otimes_R \mathscr{D} \approx \mathscr{D}/P_i\mathscr{D}.$$

Since $\mathscr{D}/P_i\mathscr{D}$ is simple, J_i/J_{i+1} is either zero or isomorphic to $\mathscr{D}/P_i\mathscr{D}$. □

3.5 COROLLARY r.gl.dim $\mathscr{D}$ = r.gl.dim R.

Proof It suffices to show that $pdI_{\mathscr{D}} <$ r.gl.dim R for all right ideals I of $\mathscr{D}$. Clearly, $pdT\mathscr{D}_{\mathscr{D}} \leq pdT_R$ for all ideals T of R since $\mathscr{D}_R$ is free. By repeated applications of 3.4 we have $pdI_{\mathscr{D}} \leq \max\{pdI_{\mathscr{D}}^{**}, \text{r.gl.dim } R - 1\}$. Thus, it suffices to show that

$$pdI_{\mathscr{D}}^{**} \leq \text{r.gl.dim } R - 1.$$

Since $0 \to K \to \mathscr{D}^n \xrightarrow{f} I^* \to 0$ is exact for some $n > 0$, so are

$$0 \to I^{**} \to \mathscr{D}^{n*} \to C = \operatorname{coker} f^* \to 0$$

and

$$0 \to C^* \to \mathscr{D}^{n**} \to I^{***} \to C' = \operatorname{coker} f^{**} \to 0.$$

However, by a diagram chase, one obtains that $K \approx C^*$ and hence, $\mathscr{D}^{n*}/I^{**} \hookrightarrow C^{**}$ which is clearly a submodule of a free $\mathscr{D}$-module. Thus, $pdI_{\mathscr{D}}^{**} = 0$ or $\leq$ r.gl.dim $\mathscr{D} - 2$ which implies $pdI_{\mathscr{D}}^{**} \leq$ r.gl.dim $R - 1$ since r.gl.dim $\mathscr{D} \leq$ r.gl.dim $R + 1$.

Remark In the particular case we are considering, since r.gl.dim $\mathscr{D} \leq 2$, projectivity of I^{**} follows immediately from 4.16. □

Next, we shall show that each right ideal I of $\mathscr{D}$ can be expressed in the form $I = \sum_{i=0}^{m} d_i\mathscr{D}$ where, for each $0 \leq k \leq m - 1$, there exists an irreducible $p_k \in R$ such that $d_{k+1}p_k \in \sum_{i=0}^{k} d_i\mathscr{D}$ and

$$\sum_{i=0}^{k+1} d_i\mathscr{D} \Big/ \sum_{i=0}^{k} d_i\mathscr{D} \approx \mathscr{D}/p_k\mathscr{D}.$$

By 2.33, $I/d_0\mathscr{D}$ is Artinian whenever d_0 is any element of I. Using the fact that if $J = d_1\mathscr{D} + d_0\mathscr{D}$, $d_1N \subseteq d_0\mathscr{D}$ for some ideal of R (see the proof of Theorem 3.4), the result follows immediately.

Let I be any right ideal of $\mathscr{D}$. Then

$$\mathscr{D} \oplus I \approx \mathscr{D} \oplus \mathscr{D}. \tag{2}$$

(Compare this with Proposition 2.27.)

Note that while $\mathscr{D}$ is not a pri-domain (for example, $I = x\mathscr{D} + \delta\mathscr{D}$ is not principal), it satisfies a property much stronger than 2.28.

To show (2) is satisfied, note that since

$$\sum_{i=0}^{k+1} d_i\mathscr{D} \Big/ \sum_{i=0}^{k} d_i\mathscr{D} \approx \mathscr{D}/p_k\mathscr{D}$$

Schanuel's lemma yields

$$\mathscr{D} \oplus \sum_{i=0}^{k} d_i\mathscr{D} \approx p_k\mathscr{D} \oplus \sum_{i=0}^{k+1} d_i\mathscr{D}$$

$$\approx \mathscr{D} \oplus \sum_{i=0}^{k+1} d_i\mathscr{D}.$$

Hence, by decreasing induction, $\mathscr{D} \oplus I \approx \mathscr{D} \oplus \mathscr{D}$. Another important consequence of this is the following: since $\mathscr{D} \oplus I \approx \mathscr{D} \oplus \mathscr{D}$ for all right ideals I of $\mathscr{D}$, applying 2.36 we obtain that each projective module of $\mathscr{D}$ is either free or isomorphic to a right ideal of $\mathscr{D}$. Thus, $M_n(\mathscr{D})$ is principal $n \geq 2$ (Chevalley [36], Th. 7, p. 20). However, since $\mathscr{D}$ is not, we obtain an example showing that being a pri (resp. pli)-ring is *not* preserved under similarity. We now summarize what we have obtained:

3.6 THEOREM *Let $R = k[X]$ where k is a field and* char $k = 0$, *δ the ordinary formal derivative of R, and $\mathscr{D}$, the ring of differential polynomials with coefficients in R. Then $\mathscr{D}$ has the following properties:*

(*a*) *$\mathscr{D}$ is a simple two-sided Noetherian domain;*

(*b*) r.gl.dim $\mathscr{D} = 1$;

(*c*) *$\mathscr{D}$ is not a pri (pli)-domain;*

(*d*) *$M_n(\mathscr{D})$ is a pri (pli)-ring $\forall\, n \geq 2$;*

(*e*) $\mathscr{D} \oplus I \approx \mathscr{D} \oplus \mathscr{D}$, $\forall I \in \text{lat}\, \mathscr{D}_{\mathscr{D}}$;

(*f*) *each finitely generated projective $\mathscr{D}$-module $P \approx \mathscr{D}^{n-1} \oplus I$, where I is isomorphic to a right ideal of $\mathscr{D}$ and $n \geq 1$.*

Generalities on differential modules

Returning to the general case, let R be an ordinary differential ring, assumed commutative, and $\mathscr{D} = \mathscr{D}_R$. Any $M \in \mathscr{D}$-mod is usually called a **differential module**. Note that such an M is necessarily an (R, R)-bimodule and a $(\mathscr{D}, R)$-bimodule as well. The action of δ on an element $m \in M$ is written $\delta(m)$. If $M, N \in \mathscr{D}$-mod, $M \otimes_R N$ and $\hom_R(M, N)$ are in $\mathscr{D}$-mod, that is, are differential modules. To see this, for $m \in M$, $n \in N$, define

$$r(m \otimes n) = rm \otimes n \quad \forall r \in R$$

and $\delta(m \otimes n) = \delta(m) \otimes n + m \otimes \delta(n)$. That $M \otimes_R N \in \mathscr{D}$-mod is easy to check by linearity.

For $f \in \hom_R(M, N)$, define

$$rf(m) = r(f(m)) \quad \forall r \in R, m \in M$$

and $\delta \circ f(m) = \delta(f(m)) - f(\delta(m))$, $\forall m \in M$.

It is also easy to check that $\delta \circ f \in \hom_R(M, N)$ and $\hom_R(M, N) \in \mathscr{D}$-mod. As an immediate consequence of the fact that $M \otimes_R N$ and $\hom_R(M, N)$ are differential modules, we obtain

3.7 PROPOSITION *For all $M, N, P \in \mathscr{D}$-mod,*

$$\hom_{\mathscr{D}}(M \otimes_R N, P) \approx \hom_{\mathscr{D}}(M, \hom_R(N, P)).$$

Proof Let $\phi : \hom_R(M \otimes_R N, P) \to \hom_R(M, \hom_R(N, P))$ be the canonical isomorphism. (Cf. Cartan–Eilenberg [56].) One simply checks that if $f \in \hom_{\mathscr{D}}(M \otimes {}_R N, P)$ then $\phi(f) \in \hom_{\mathscr{D}}(M, \hom_R(N, P))$. This is routine and left for the reader. □

The next proposition will find use in the next chapter. Recall that $M \in$ mod-R is **reflexive** (**torsionless**) if the canonical map of M into M^{**} is an isomorphism (monomorphism).

3.8 PROPOSITION *Let $M \in \mathscr{D}$-mod be reflexive in* mod-R. *Then $M \overset{\text{nat}}{\approx} M^{**}$ as $\mathscr{D}$-modules.*

Proof Since $M \overset{\text{nat}}{\approx} M^{**}$ as R-modules by hypothesis, all that remains is to check that the natural isomorphism above is an isomorphism of $\mathscr{D}$-modules. This is left for the reader. □

It is possible to generalize the notion of the ring of differential polynomials in an obvious manner. For, suppose R is a **partial differential ring**, that is to say, R is a ring with n commuting derivations, $\delta_1, \ldots, \delta_n$ defined on R. Define $\mathscr{D}$ to be all polynomials in the δ_i with coefficients in R. A typical element of $\mathscr{D}$ has the form

$$\sum_{(\nu_1,\ldots,\nu_n)\in N^n} a_{\nu_1\ldots\nu_n}\delta_1{}^{\nu_1}\ldots\delta_n{}^{\nu_n}, \quad a_{\nu_1\ldots\nu_n} \in R.$$

Equality and addition are defined as in commuting polynomial rings, with multiplication being induced by the relations

$$\delta_i a = a\delta_i + \delta_i(a) \quad \forall a \in R, i = 1, \ldots, n.$$

$\mathscr{D}$ is also called the **ring of linear differential operators with coefficients in** R.

This ring can be inductively defined as the differential polynomial ring with respect to the derivation δ_n over the differential polynomial ring with respect to the derivations $\delta_1, \ldots, \delta_{n-1}$ over R. Thus, by 3.1, $\mathscr{D}$ is Noetherian whenever R is.

Twisted group rings

Let R be a ring and G a group acting on R. Recall that the **twisted group ring** $L_G(R)$ over R (with a group G acting on R) can be defined as follows:

$$(L_G(R), +) = (R(G), +),$$

where $R(G)$ is the ordinary group ring; multiplication in $L_G(R)$ is induced by the relations

$$\sigma a = \sigma(a)\sigma \quad \forall \sigma \in G, a \in R.$$

(Here $\sigma(a)$ represents the group action of $\sigma \in G$ on $a \in R$.)

It is well known that for $G = \langle \rho \rangle$ (the infinite cyclic group generated by ρ), $L_G(R) = L_\rho(R)$ is a simple pli (pri)-domain whenever R is a field and ρ is an automorphism of R having infinite period (Jacobson [43], p. 38 and Jacobson [64*a*], p. 211). Moreover, in this case, each left (right) ideal of $L_\rho(R)$ can be generated by an element $a \in L_\rho(R)$ having the form $a = \sum_{i=0}^{n} a_i \rho^i$. Since a fixed R and G will be assumed during the ensuing discussion, we shall replace the symbol $L_G(R)$ with L.

Let $M \in L$-mod, and let R be commutative. Note that if we denote the L action of σ on $m \in M$ by $\sigma(m)$, we obtain for each $\sigma \in G$, an additive homomorphism of M satisfying $\sigma(am) = \sigma(a)\sigma(m)$ for all $a \in R$. Using these observations, it is quite easy to define an L action on $\hom_R(M, N)$ and $M \otimes_R N$ whenever $M, N \in L$-mod. $\hom_R(M, N)$: for $f \in \hom_R(M, N)$, $\sum_\sigma a_\sigma \sigma \in L$, define

$$\sum_\sigma a_\sigma \sigma f(m) = \sum_\sigma a_\sigma \sigma(f(\sigma^{-1}(m))) \quad \forall m \in M.$$

$M \otimes_R N$: for $\sum_i m_i \otimes n_i \in M \otimes_R N$, $\sum_\sigma a_\sigma \sigma \in L$, define

$$\sum_\sigma a_\sigma \sigma \sum_i m_i \otimes n_i = \sum_{\sigma, i} a_\sigma(\sigma(m_i) \otimes \sigma(n_i)).$$

One readily verifies that the above definitions actually do endow $\hom_R(M, N)$ and $M \otimes_R N$ with the structure of an L-module.

3.9 DEFINITION For $M \in L$-mod, the set $I(M) = \{m \in M \mid \sigma(m) = m, \sigma \in G\}$ will be called the G-**invariant subset** of M. Other notation, M^G.

Remark In general, $I(M)$ is not an L-submodule of M, merely an $I(L)$ submodule of M.

The proofs of the following propositions are, for the most part, routine, and will therefore be omitted.

3.10 PROPOSITION *Let $M, N \in L$-mod. Then*

$$I(\hom_R(M, N)) = \hom_L(M, N). \quad \square$$

R is naturally a left L-module, namely, for all $a = \sum_\sigma a_\sigma \sigma \in L$, $r \in R$,

$$ar = \sum_\sigma a_\sigma \sigma(r).$$

We shall always denote this L-module by ${}_LR$.

3.11 COROLLARY *For $M \in L$-mod, $I(M) = \hom_L(R, M)$.* $\square$

3.12 PROPOSITION *For all $M, N, P \in L$-mod, the natural isomorphism*

$$\hom_R(M \otimes {}_RN, P) \approx \hom_R(M, \hom_R(N, P))$$

is L-linear. $\square$

3.13 COROLLARY

$$\hom_L(M \otimes {}_RN, P) \approx \hom_L(M, \hom_R(N, P)). \quad \square$$

3.14 PROPOSITION *For $M \in L$-mod, M reflexive in* Mod-R, *the natural isomorphism*

$$M \overset{\Psi}{\approx} \hom_R(\hom_R(M, R), R) = M^{**}$$

is an isomorphism of L-modules. $\square$

Skew polynomial rings†

Let ρ be an endomorphism of a field k. A map $\delta : k \to k$ is called a ρ-**derivation** of k if $\delta(a + b) = \delta(a) + \delta(b)$ and $\delta(ab) = \rho(a)\delta(b) + \delta(a)b$ hold for all $a, b \in k$. Using k, ρ and δ one can define the ring $R = k[t; \rho, \delta]$ of skew polynomials in t, consisting of all polynomials

$$\sum_{i=0}^{n} a_i t^i,$$

† This section (Cozzens [72]) is reproduced from *Journal of Algebra*, Volume 23, October 1972, by permission of Academic Press, New York.

where $a_i \in k$ for each i, with the usual addition, multiplication being defined by

$$ta = \rho(a)t + \delta(a) \ \forall\ a \in k.$$

It is not difficult to verify that $R = k[t; \rho, \delta]$ is a pli-domain and is principal on the right if and only if ρ is surjective. Since $k[t; \rho, \delta]$ is never simple if $\delta = 0$ and $k[t; \rho, \delta]$ is usually simple if $\rho = id$, one naturally asks whether $k[t; \rho, \delta]$ can be simple if ρ is *not* surjective and δ is a *nontrivial* ρ-derivation.

If k is commutative, the answer is no. For by a lemma due to Cohn ([61*a*], Lemma page 537), if k is commutative, then any ρ-derivation δ of k is inner [i.e., for some $x \in k$, $\delta(a) = \rho(a)x - xa$ for all $a \in k$] unless ρ is an inner automorphism of k. If δ is an inner ρ-derivation of k determined by $x \in k$, then one easily checks that the principal left ideal of $R = k[t; \rho, \delta]$, $R(t + x)$, is two-sided and nontrivial. Thus, one is naturally forced to consider noncommutative k.

Theorem 3.15 provides complicated sufficient conditions which lead to simple pli-domains. Finding a triple (k, ρ, δ) satisfying the hypothesis of 3.15 is where the fun begins. A remarkable construction due to Cohn [61*a*] and an embedding theorem due to Jategaonkar [69] ultimately produce (k, ρ, δ).

3.15 THEOREM *Let E be a field of characteristic 0, $\rho : E \to E$ a ring monomorphism and δ a ρ-derivation of E satisfying*

(*a*) $\rho\delta + \delta\rho = 0$.

(*b*) *If $x \in E$ satisfies $x\rho^n(\alpha) = \rho^{n-j}(\alpha)x$ for all $\alpha \in E$ where $n \geq j \geq 1$ are both arbitrary, then $x = 0$.*

(*c*) *There does not exist an $x \in E$ such that for some $m \in \mathbb{N}$ and for all $\alpha \in E$*

$$\delta^n[\rho^m(\alpha)] + x\rho^m(\alpha) = \rho^{m+n}(\alpha)x \quad n = 1, 2.$$

(*d*) *There does not exist an $x \in E$ satisfying $\rho(x) = x$ and $\delta(x) = 0$ such that for some $n \in \mathbb{N}$ and for all $\alpha \in E$,*

$$\delta^n(\alpha) + x\alpha = \rho^n(\alpha)x.$$

Then $R = E[t; \rho, \delta]$ is a simple pli-domain.

Proof Let I be an ideal of R such that $0 \subset I \subset R$. Then

$I = Rf$ for some $f \in R$, where $f \neq 0$. The proof will be by contradiction.

Let $f = \sum_{i=0}^{m} a_i t^i$, where we can assume that $a_m = 1$.

Case 1. $m = 2n$.

Given $\alpha \in E$,

$$f\alpha = \beta f \quad \text{for some } \beta \in E, \tag{3}$$

since I is an ideal. Expanding Eq. (3) and equating coefficients we obtain

$$\begin{aligned} \rho^{2n}(\alpha) &= \beta, \\ a_{2n-1}\rho^{2n-1}(\alpha) &= \beta a_{2n-1}, \\ a_{2n-1}\delta\rho^{2n-2}(\alpha) + a_{2n-2}\rho^{2n-2}(\alpha) &= \beta a_{2n-2}, \\ &\vdots \\ \sum_{i=0}^{2n} a_i \delta^i(\alpha) &= \rho^{2n}(\alpha)a_0. \end{aligned} \tag{4}$$

By successive applications of (b) and (c), we obtain $a_{2n-1} = a_{2n-2} = \ldots = a_1 = 0$. Now

$$ft = (t + \beta')f \quad \text{for some } \beta' \in E,$$

i.e., $(t^{2n} + a_0)t = (t + \beta')(t^{2n} + a_0)$. Hence

$$t^{2n+1} + a_0 t = t^{2n+1} + \rho(a_0)t + \delta(a_0) + \beta' t^{2n} + \beta' a_0.$$

Thus $\beta' = 0$ and, consequently,

$$a_0 = \rho(a_0) \quad \text{and} \quad \delta(a_0) = 0.$$

By (4) $\delta^{2n}(\alpha) + a_0\alpha = \rho^{2n}(\alpha)a_0$ for all $\alpha \in E$ contradicting (d). Thus $f = t^{2n}$. Since

$$t^{2n}\alpha = \sum_{j=0}^{n} \binom{n}{j} \rho^{2n-2j}[\delta^{2j}(\alpha)]t^{2n-2j},$$

$\rho^{2n-2}(\delta^2(\alpha)) = 0$ for all $\alpha \in E$. Therefore, $\delta^2(\alpha) = 0$ for all $\alpha \in E$, contradicting (c).

Case 2. $m = 2n + 1$.

Proceeding as above, we obtain

$$\begin{aligned} \rho^{2n+1}(\alpha) &= \beta, \\ \delta(\rho^{2n}(\alpha)) + a_{2n}\rho^{2n}(\alpha) &= \beta a_{2n}, \\ &\vdots \\ \sum_{i=0}^{2n+1} a_i\delta^i(\alpha) &= \beta a_0. \end{aligned} \tag{5}$$

Clearly, Eq. (5) contradicts (c).

Thus, no such f can exist, implying that $I = 0$ or $I = R$. □

Definition of E, ρ and δ

Let k be any commutative field of characteristic 0 and denote by A, the free associative algebra over k on the free generating set $B = \{x_0, x_1, \ldots\}$. Since A is the free associative algebra on B, there exists a unique endomorphism ρ of A such that

$$\rho(x_0) = -x_0 \quad \text{and} \quad \rho(x_i) = x_{i+1} \quad \forall i \geq 1.$$

Clearly ρ is injective. We define

$$[x_i, x_0] = \rho(x_i)x_0 - x_0x_i \quad \forall i \geq 1.$$

Let $\bar{k} = k(t_{in}, i \in \mathbb{N} \cup \{0\}, n \in \mathbb{N})$ be the rational function field in the indeterminates t_{in} over k. Let σ be the k-endomorphism of $\bar{k}$ defined by $\sigma(t_{in}) = t_{i,n+1}$ for all $i \in \mathbb{N} \cup \{0\}$ and $n \in \mathbb{N}$.

Let $D = \bar{k}[Y, \sigma]$ be the twisted polynomial ring in Y with coefficients in $\bar{k}$. Let K be the left quotient field of D. If we set

$$\hat{x}_i = t_{i1}Y$$

then $\hat{x} : A \to K$ is an embedding (see Jategaonkar [69]). Identify x_i with $\hat{x}_i \in K$. We shall extend ρ to K as follows:

Define

$$\begin{aligned} \rho(t_{on}) &= -t_{on} \quad \forall n \in \mathbb{N}, \\ \rho(t_{in}) &= t_{i+1,n} \quad \forall i \geq 1, \end{aligned}$$

and

$$\rho(Y) = Y$$

Clearly, $\rho\sigma = \sigma\rho$ and

$$\begin{aligned} \rho(t_{o1}Y) &= -t_{o1}Y \\ \rho(t_{i,1}Y) &= t_{i+1,1}Y \quad \forall i \geq 1. \end{aligned}$$

Since $\rho\sigma = \sigma\rho$, ρ extends to a ring monomorphism of K and moreover, does what it should with respect to the generators of A.

Let E be the subfield of K generated over k by x_i for all $i \geq 1$ and $[x_i, x_o]$ for all $i \geq 1$. Define a ρ-derivation δ of K as follows:

$$\delta(\alpha) = \rho(\alpha)x_o - x_o\alpha \quad \forall\alpha \in K.$$

Clearly, $\rho\delta + \delta\rho = 0$ and E admits both ρ and δ. Also note that x_o and $x_o^2 \notin E$. (See Cohn [61*a*].) The following lemmas show that (E, ρ, δ) satisfies the hypothesis of Theorem 3.15.

Remark $\bar{k}^\sigma = \{\alpha \in \bar{k} \mid \sigma(\alpha) = \alpha\} = k$ and

$$\bar{k}^\rho = k(t_{o1}^2, t_{o2}^2, \ldots, t_{on}^2, \ldots).$$

3.16 LEMMA *If $a \in K$ and $at_{n1}Y = t_{n-j,1}Ya$ where $n - j \geq 0$, for all $n \geq N > 0$, N some suitably large integer, then $j = 0$ and $a \in k$.*

Proof Set $t_{n1}Y = t_nY$ and assume that $a \in K$ is of the form

$$a = \left(\sum_{i=0}^{m^1} a_iY^i\right) \Big/ \left(\sum_{i=0}^{m} b_iY^i\right).$$

Now,

$$at_nY = \left(\sum_{i=0}^{m^1} a_i\sigma^i(t_n)Y^{i+1}\right) \Big/ \left(\sum_{i=0}^{m} b_iY^i\right)$$

and

$$\left(\sum_{i=0}^{m} c_iY^i\right)t_{n-j}Y = t_nY\left(\sum_{i=0}^{m} b_iY^i\right),$$

where

$$c_i = t_n\sigma(b_i)/\sigma^i(t_{n-j}) \quad \text{for} \quad 0 \leq i \leq m.$$

Therefore,

$$t_{n-j}Ya = \left(\sum_{i=0}^{m^1} t_n\sigma(a_i)Y^i\right) \Big/ \left(\sum_{i=0}^{m} c_iY^i\right)$$

by definition of multiplication in K.

If

$$\left(\sum_{i=0}^{k} d_iY^i\right)\left(\sum_{i=0}^{m} c_iY^i\right) = \left(\sum_{i=0}^{k} e_iY^i\right)\left(\sum_{i=0}^{m} b_iY^i\right), \tag{6}$$

then, in particular,

$$d_k\sigma^k(c_m) = e_k\sigma^k(b_m).$$

Hence,

$$\left(\sum_{i=0}^{k} d_i Y^i\right)\left(\sum_{i=0}^{m^1} t_n\sigma(a_i) Y^i\right) = \left(\sum_{i=0}^{k} e_i Y^i\right)\left(\sum_{i=0}^{m^1} a_i\sigma^i(t_n) Y^i\right). \quad (7)$$

Thus,

$$d_k\sigma^k[t_n\sigma(a_{m^1})] = e_k\sigma^k[a_{m^1}\sigma^{m^1}(t_n)],$$

thus,

$$a_{m^1}\sigma(b_m)/\sigma(a_{m^1})b_m = \sigma^m(t_{n-j})/\sigma^{m^1}(t_n) \quad \forall n \geq N.$$

If $m \neq m^1$ or $j > 0$, let $n \to \infty$ to get the desired contradiction. Hence, $m = m^1$ and $j = 0$. Now $c_o = \sigma(b_o)$, $d_oc_o = e_ob_o$ and $d_ot_n\sigma(a_o) = e_oa_ot_n$. Therefore,

$$\sigma(a_o)/a_o = e_o/d_o = c_o/b_o = \sigma(b_o)/b_o,$$

implying that $\sigma(a_o/b_o) = a_o/b_o$. Thus,

$$a_o = \alpha b_o \quad \text{where} \quad \alpha \in k.$$

Let $t = t_n$.

$$d_oc_1 + d_1\sigma(c_o) = e_ob_1 + e_1\sigma(b_o) \quad (8)$$

and

$$d_ot\sigma(a_1) + d_1\sigma(t)\sigma^2(a_o) = e_oa_1\sigma(t) + e_1\sigma(a_o)\sigma(t). \quad (9)$$

Now

$$c_1 = t\sigma(b_1)/\sigma(t) \quad \text{and} \quad \sigma(b_o) = c_o.$$

By substituting these expressions in (8) we obtain

$$d_ot\sigma(b_1)/\sigma(t) + d_1\sigma^2(b_o) = e_ob_1 + e_1\sigma(b_o).$$

Since $a_o = \alpha b_o$, we obtain

$$d_ot\sigma(\alpha b_1) + d_1\sigma(t)\sigma^2(a_o) = e_o\sigma(t)b_1 + e_1\sigma(t)\sigma(a_o). \quad (10)$$

Comparing (9) and (10), we obtain

$$t(\sigma(a_1) - \sigma(\alpha b_1)) = (\sigma(b_o)/b_o)\sigma(t)(a_1 - \alpha b_1). \quad (11)$$

Equation (11) holds for all $n \geq N$ (remember, $t = t_n$). If $a_1 - \alpha b_1 \neq 0$, letting $n \to \infty$ produces a contradiction.

Assume $a_i = \alpha b_i$ for all $i \leq m - 1$ by induction. By a process similar to the above (by examining the coefficient of Y^{m+k-1} in (6) and (7)), we obtain

$$\begin{aligned}\sigma^{m+k-1}(t)e_{k-1}\sigma^{k-1}(\alpha b_m) &+ \sigma^{m+k-1}(t)e_k\sigma^k(\alpha b_{m-1}) \\ &- d_k\sigma^k(t)\sigma^{k+1}(\sigma b_{m-1}) \\ = \sigma^{m+k-1}(t)e_{k-1}\sigma^{k-1}(a_m) &+ \sigma^{m+k-1}(t)e_k\sigma^k(a_{m-1}) \\ &- d_k\sigma^k(t)\sigma^{k+1}(a_{m-1}).\end{aligned}$$

Thus, $a_m = \alpha b_m$ and hence,

$$a = \alpha \in k. \qquad \square$$

3.17 Lemma *If $a \in E$ satisfies $\rho(a) = a$ and $\delta(a) = 0$, then $a \in k$.*

Proof We can assume once again that

$$a = \left(\sum_{i=0}^{n} a_i Y^i\right) \Big/ \left(\sum_{i=0}^{m} b_i Y^i\right), \quad b_m = 1.$$

$\delta(a) = 0$ implies that $t_{01}\sigma(a) = at_{01}$. Let $t = t_{01}$,

$$\sigma(a) = \left(\sum_{i=0}^{n} \sigma(a_i) Y^i\right) \Big/ \left(\sum_{i=0}^{m} \sigma(b_i) Y^i\right)$$

and

$$at = \left(\sum_{i=0}^{n} a_i\sigma^i(t) Y^i\right) \Big/ \left(\sum_{i=0}^{m} b_i Y^i\right).$$

$$\left(\sum_{i=0}^{m} c_i Y^i\right) t = t\left(\sum_{i=0}^{m} \sigma(b_i) Y^i\right),$$

where $c_i = t\sigma(b_i)/\sigma^i(t)$ for all $0 \leq i \leq m$.

Thus,

$$\left(\sum_{i=0}^{n} t\sigma(a_i) Y^i\right) \Big/ \left(\sum_{i=0}^{m} c_i Y^i\right) = \left(\sum_{i=0}^{n} a_i\sigma^i(t) Y^i\right) \Big/ \left(\sum_{i=0}^{m} b_i Y^i\right).$$

If

$$\left(d = \sum_{i=0}^{k} d_i Y^i\right)\left(\sum_{i=0}^{m} c_i Y^i\right) = \left(e = \sum_{i=0}^{k} e_i Y^i\right)\left(\sum_{i=0}^{m} b_i Y^i\right)$$

then, in particular, $d_k\sigma^k(c_m) = e_k$ and

$$d_k\sigma^k(t\sigma(a_n)) = e_k(\sigma^k(a_n\sigma^n(t)))$$

since

$$d\left(\sum_{i=0}^{n} t\sigma(a_i)Y^i\right) = e\left(\sum_{i=0}^{n} a_i\sigma^i(t)Y^i\right)$$

by definition of equality in K. Therefore,

$$t\sigma(a_n)/a_n\sigma^n(t) = t/\sigma^m(t)$$

implying that

$$\sigma(a_n)/a_n = \sigma^n(t)/\sigma^m(t). \tag{12}$$

Now,

$$\left(\sum_{i=0}^{n} \rho(a_i)Y^i\right)\Big/\left(\sum_{i=0}^{m} \rho(b_i)Y^i\right) = \left(\sum_{i=0}^{n} a_iY^i\right)\Big/\left(\sum_{i=0}^{m} b_iY^i\right)$$

since $\rho(a) = a$. Therefore if

$$\left(\sum_{i=0}^{k} f_iY^i\right)\left(\sum_{i=0}^{m} \rho(b_i)Y^i\right) = \left(\sum_{i=0}^{k} g_iY^i\right)\left(\sum_{i=0}^{m} b_iY^i\right),$$

$$f_k\sigma^k(\rho(b_m)) = g_kb_m$$

or $f_k = g_k$ since $b_m = 1$. Thus, $\rho(a_n) = a_n$ by definition of equality in K.

By (12) and the definition of ρ, $m = n$ and $\sigma(a_n) = a_n$. Thus, $a_n = \alpha \in k$.

Now,

$$d_{k-1}\sigma^{k-1}(c_m) + d_k\sigma^k(c_{m-1}) = e_{k-1}\sigma^{k-1}(b_m) + e_k\sigma^k(b_{m-1})$$

and

$$d_{k-1}\sigma^{k-1}(t\sigma(a_m)) + d_k\sigma^k(t\sigma(a_{m-1}))$$
$$= e_{k-1}\sigma^{k-1}(a_m\sigma^m(t)) + e_k\sigma^k(a_{m-1}\sigma^{m-1}(t)).$$

Therefore,

$$e_k\sigma^{k+m-1}(t)[\sigma^k(\alpha b_{m-1} - a_{m-1})] = d_k\sigma^k(t)[\sigma^{k+1}(\alpha b_{m-1} - a_{m-1})],$$
$$e_k = d_k\sigma^k(c_m).$$

Hence

$$c_m\sigma^{m-1}(t)[\alpha b_{m-1} - a_{m-1}] = t\sigma(\alpha b_{m-1} - a_{m-1})$$

or assuming that $\alpha b_{m-1} - a_{m-1} \neq 0$,

$$t\sigma^{m-1}(t)/\sigma^m(t)t = \sigma(\alpha b_{m-1} - a_{m-1})/(\alpha b_{m-1} - a_{m-1}). \tag{13}$$

$$f_{k-1}\sigma^{k-1}(\rho(b_m)) + f_k\sigma^k(\rho(b_{m-1})) = g_{k-1}\sigma^{k-1}(b_m) + g_k\sigma^k(b_{m-1}) \tag{14}$$

and

$$f_{k-1}\sigma^{k-1}(\rho(a_m)) + f_k\sigma^k(\rho(a_{m-1})) = g_{k-1}\sigma^{k-1}(a_m) + g_k\sigma^k(a_{m-1}).$$

Therefore,

$$f_k\sigma^k(\rho(\alpha b_{m-1} - a_{m-1})) = g_k\sigma^k(\alpha b_{m-1} - a_{m-1})$$

and $f_k = g_k$ imply that

$$\rho(\alpha b_{m-1} - a_{m-1}) = \alpha b_{m-1} - a_{m-1} = \beta.$$

Thus, $\beta \in \bar{k}^\rho$. By (1), $\sigma(\beta)\sigma[\sigma^{m-1}(t)] = \beta\sigma^{m-1}(t)$. Therefore,

$$\beta\sigma^{m-1}(t) = \gamma \in \bar{k}^\sigma = k.$$

Thus $a_{m-1} = \alpha b_{m-1}$ since $\rho(\beta) = \beta$ and $\rho(\gamma) = \gamma$.

Assume by induction that $a_j = \alpha b_j$ for all $1 \leq j \leq m$.

$$\sum_{i=0}^{k} d_i\sigma^i(c_{n-i}) = \sum_{i=0}^{k} e_i\sigma^i(b_{n-i})$$

and

$$\sum_{i=0}^{k} d_i\sigma^i(t\sigma(a_{n-i})) = \sum_{i=0}^{k} e_i\sigma^i(a_{n-i}\sigma^{n-i}(t)).$$

Consequently,

$$d_k[\sigma^{k+1}(a_o - \alpha b_o)] = e_k[\sigma^k(a_o - \alpha b_o)]$$

by induction. As before, $\rho(a_o - \alpha b_o) = (a_o - \alpha b_o)$ and hence $a_o = \alpha b_o$. Therefore, $a = \alpha \in k$. □

3.18 Lemma (E, ρ, δ) *satisfies the hypothesis of Theorem* 3.15.

Proof (*a*) $\rho\delta + \delta\rho = 0$ is trivially satisfied by construction.
(*b*) If there exists an $x \in E$ such that $\forall \alpha \in E$,

$$x\rho^n(\alpha) = \rho^{n-j}(\alpha)x \quad n \geq j \geq 1$$

then in particular,

$$x\rho^n(x_i) = xx_{n+i} = xt_{n+i}Y = \rho^{n-j}(x_i)x = t_{n-j+i,1}Yx,$$

$\forall i > 0$, contradicting 3.16.

(*c*) Note that $\delta(\alpha) = \rho(\alpha)x_o - x_o\alpha$ and $\delta^2(\alpha) = \rho^2(\alpha)x_o^2 - x_o^2\alpha$, $\forall\alpha \in E$ by definition of δ and δ^2. If

$$\delta[\rho^m(\alpha)] = \rho^{m+1}(\alpha)x - x\rho^m(x) \quad \forall\alpha \in E,$$

where $x \in E$, then

$$\rho^{m+1}(\alpha)(x_o - x) = (x_o - x)\rho^m(\alpha)$$

$\forall\alpha \in E$ contradicting (*b*).

A similar argument works for the $n = 2$ case.

(*d*) If there exists an $x \in E$ satisfying $\rho(x) = x$ and $\delta(x)$ such that for some $n \in \mathbb{N}$ and for all $\alpha \in E$,

$$\delta^n(\alpha) + x\alpha = \rho^n(\alpha)x,$$

then by 3.17, $x \in k$. Thus,

$$\delta^n(\alpha) = x[\rho^n(\alpha) - \alpha]$$

which just cannot happen. □

Jacobson domains

We complete this chapter with yet another class of simple domains which we call Jacobson domains. If D is any field with center C, and if the polynomial ring $D[x]$ is primitive, then $D \otimes_C C(x)$ is a simple pli- and pri-domain, not a field. This holds, for example, whenever D is transcendental over C, that is, whenever there exists $d \in D$ which is not algebraic over C.

The first proposition is taken from Jacobson [43].

3.19 Proposition *If D is a field with center C, then every ideal I of the polynomial ring $D[x]$ is generated by a central polynomial $p(x) \in C[x]$.*

Proof Since $R = D[x]$ has a right and left division algorithm R is a principal (right and left) ideal domain. Thus $I = p(x)R$, for some monic $p(x) \in R$ of least degree. Then, $\forall d \in D$, $r(x) = dp(x) - p(x)d$ is a polynomial in I of lower degree. It follows that $r(x) = 0$, and so $p(x)$ commutes with every $d \in D$, hence $p(x)$ lies in the center $C[x]$ of R. □

3.20 Theorem (Jacobson [64*a*]) *If D is a field, then $D[x]$ is not primitive if and only if every nonzero $f(x) \in D[x]$ is a factor* *of a nonzero central polynomial, that is, iff* $\forall f(x) \neq 0$ *there exist* $g(x) \in D[x]$ *and* $0 \neq p(x) \in C[x]$ *with* $f(x)g(x) = p(x)$.

Proof $R = D[x]$ is primitive iff there exists a simple faithful right module V, or equivalently some maximal right ideal $m(x)R$ contains no nonzero ideal. Using the last proposition we see that R is not primitive iff every maximal right ideal $m(x)R$ contains a nonzero central polynomial $p(x)$ that is, iff $m(x)$ is a factor of a nonzero central polynomial. However, this is equivalent to stating that every nonzero $f(x)$ is a factor of a nonzero central polynomial, since $f(x)$ a product of irreducible polynomials $m_1(x), \ldots, m_t(x)$, and every irreducible polynomial $m(x)$ generates a maximal right ideal. $\square$

In the next theorem, $C(x)$ is the field of rational functions over C.

3.21 THEOREM *Let* D *be a field with center* C. *Then* $A = D \otimes_C C(x)$ *is a simple pli- and pri-domain embedded in the right quotient field* Q *of* $D[x]$. *Moreover* $A = Q$ *iff* $D[x]$ *is not primitive. In particular, whenever* D *is transcendental over* C, *then* A *is not a field.*

Proof We know that Q exists since $D[x]$ is a right Ore domain. (Any polynomial ring over an Ore domain is an Ore domain.) Furthermore A embeds in Q under the mapping

$$f(x) \otimes p(x)^{-1} \mapsto f(x)p(x)^{-1}$$

defined for all $f(x) \in D[x]$, $p(x) \in C[x]$. If $A = Q$, and if $f(x) \in D[x]$, then

$$f(x)^{-1} = g(x)p(x)^{-1}$$

for suitable $g(x) \in D[x]$ and $p(x) \in C[x]$, so

$$f(x)g(x) = p(x)$$

is a factor of a central polynomial. Conversely, if the latter holds for any $f(x) \in D[x]$, then A contains $f(x)^{-1} = g(x)p(x)^{-1}$ for any $f(x) \neq 0$, so that $A \supseteq Q$, and hence, $A = Q$.

We complete the proof by showing that $D[x]$ is primitive whenever D is transcendental. Let $t \in D$, and suppose that

$$t(x + 1)h(x) = p(x)$$

is a factor of a central polynomial $p(x) \neq 0$. Write

$$p(x) = \alpha_0 + \alpha_1 x + \ldots + \alpha_n x^n,$$
$$h(x) = d_0 + d_1 x + \ldots + d_m x^m,$$

where $\alpha_i \in C$, $d_j \in D$, $\forall i, j$. Then, $m = n - 1$, and comparing coefficients, we see that

$$\begin{aligned} \alpha_0 &= d_0, \\ d_0 t + d_1 &= \alpha_1, \\ d_1 t + d_2 &= \alpha_2, \\ \vdots \quad & \quad \vdots \\ d_{n-2} t + d_{n-1} &= \alpha_{n-1}, \\ d_{n-1} t &= \alpha_n. \end{aligned}$$

Then, solving recursively for d_i, $i = 0, \ldots, n - 1$, yields d_{n-1} as a polynomial of degree $n - 1$ in t with coefficients in C;

$$d_1 = -\alpha_0 t - \alpha_1, \quad d_2 = -\alpha_0 t^2 - \alpha_1 t - \alpha_2, \ldots,$$

and

$$d_{n-1} = -\alpha_0 t^{n-1} - \alpha_1 t^{n-2} - \ldots - \alpha_{n-1} = \alpha_n t^{-1}.$$

Thus,

$$\alpha_0 t^n + \alpha_1 t^{n-1} + \ldots + \alpha_{n-1} t + \alpha_n = 0$$

so that t is a zero of a nonzero central polynomial, that is, t is algebraic. Using 3.20, we see that whenever $D[x]$ is not primitive, then D must be algebraic. □

4. *Orders in simple Artin rings*

In this chapter we study rings which are right orders in simple Artin rings and explicitly determine these via the Goldie and Faith–Utumi Theorems. We also present some further results concerning representations of semiprime maximal orders. Although the rings considered are more general than simple Goldie rings, the conclusions and the techniques of proof are invaluable in the structure theory of simple Goldie rings.

Recall that a right ideal I of R is called **nil** if for all $x \in I$, $x^n = 0$ for some $n > 0$.

The identity $(ra)^{n+1} = r(ar)^n a$ for elements $a, r \in R$ shows that Ra is a nil left ideal iff aR is a nil right ideal. Thus, R has no nil right ideals $\neq 0$ if and only if R has no nil left ideals $\neq 0$.

4.1 Lemma *Let S be a ring satisfying the maximum condition for right annulets. If S has a nonzero nil right or left ideal A, then S contains a nonzero nilpotent ideal.*

Proof (Utumi [63]) By the preceding remark, we can assume A is a nil left ideal $\neq 0$. Let $0 \neq a \in A$ be such that $a^\perp$ is maximal in $\{x^\perp \mid 0 \neq x \in A\}$. If $u \in S$ is such that $ua \neq 0$, then $(ua)^n = 0$ and $(ua)^{n-1} \neq 0$ for some $n > 1$. Since $[(ua)^{n-1}]^\perp \supseteq a^\perp$, and since $ua \in A$, it follows that $[(ua)^{n-1}]^\perp = a^\perp$. Since $ua \in [(ua)^{n-1}]^\perp$, then $aua = 0$. Thus, $aSa = 0$, and then $(a)^3 = 0$, where (a) is the ideal of S generated by a. □

4.2 Lemma *Let R be any ring which satisfies the* (a.c.c.)$^\perp$ *and* (a.c.c.)$\oplus$. *If $a \in R$ there exists $n > 0$ such that $a^n R + a^{n\perp}$ is an essential right ideal.*

Proof There is an $n \geq 1$ such that $a^{n\perp} = a^{n+1\perp}$. Then $a^n R \cap a^{n\perp} = 0$. Let I be a right ideal and suppose that

$$I \cap (a^n R + a^{n\perp}) = 0.$$

Then the sum $I + a^n I + a^{2n} I + \dots$ is direct and, because R has (a.c.c.)$\oplus$, we conclude that $I = 0$. □

As an immediate consequence of 4.2 we have that if $x \in R$ satisfies $x^{\perp} = 0$, then xR is an essential right ideal of R.

4.3 DEFINITION A ring R is said to satisfy the **right Ore condition** if given $a, b \in R$, b regular, $\exists a_1, b_1 \in R$, b_1 regular, such that

$$ab_1 = ba_1.$$

It is easy but somewhat tedious to show that R has a right quotient ring if and only if R satisfies the right Ore condition. (See, for example, Jacobson [64*a*], or Faith [73*a*], Chapter 9.)

Recall that $Z(R) = \{x \in R \mid x^{\perp} \text{ is an essential right ideal of } R\}$ is a right ideal of R, called the **right singular ideal** of R. It is easy to show that $Z(R)$ is in fact, a two-sided ideal of R. The next theorem provides necessary and sufficient conditions for R to have a classical right quotient ring which is a (simple) semisimple Artin ring.

4.4 THEOREM (Goldie [58, 60], Lesieur–Croisot [59]) *A ring R has a semisimple right quotient ring if and only if R is semiprime and satisfies* (a.c.c.)$\oplus$ *and* (a.c.c.)$^{\perp}$.

Proof (Goldie [69]) We first assume that R is a semiprime ring satisfying (a.c.c.)$\oplus$ and (a.c.c.)$^{\perp}$, and prove that an essential right ideal E contains a regular element. E is not a nil ideal by 4.1, so it has an element $a_1 \neq 0$ with $a_1{}^{\perp} = a_1{}^{2\perp}$. Either $a_1{}^{\perp} \cap E = 0$ or $a_1{}^{\perp} \cap E \neq 0$. In the latter case, choose $a_2 \in a_1{}^{\perp} \cap E$ with $a_2 \neq 0$ and $a_2{}^{\perp} = a_2{}^{2\perp}$. If $a_1{}^{\perp} \cap a_2{}^{\perp} \cap E \neq 0$, then the process continues.

At the general stage we have a direct sum

$$a_1R \oplus \ldots \oplus a_kR \oplus (a_1{}^{\perp} \cap \ldots \cap a_k{}^{\perp} \cap E),$$

where $a_k \in a_1{}^{\perp} \cap \ldots \cap a_{k-1}{}^{\perp} \cap E$, $a_k \neq 0$ and $a_k{}^{\perp} = a_k{}^{2\perp}$.

The process has to stop, because R has (a.c.c.)$\oplus$. Let this happen at the kth stage. Then

$$a_1{}^{\perp} \cap \ldots \cap a_k{}^{\perp} \cap E = 0 = a_1{}^{\perp} \cap \ldots \cap a_k{}^{\perp}$$

and hence

$$(a_1{}^2 + \ldots + a_k{}^2)^{\perp} = a_1{}^{\perp} \cap \ldots \cap a_k{}^{\perp} = 0.$$

Let Z be the singular ideal of R and $z \in Z$. Then $z^nR \oplus z^{n\perp}$ is an essential right ideal for some $n > 0$ and $z^{n\perp}$ is also essential. Hence $z^n R = 0$. Thus, Z is a nil ideal and hence is zero. Set $c = a_1{}^2 + \ldots + a_k{}^2 \in E$; as $c^\perp = 0$ we deduce that cR is essential by 4.2. Hence ${}^\perp c \subseteq Z$, so that ${}^\perp c = 0$ and c is a regular element. This establishes the existence of regular elements in R.

Suppose that $a, d \in R$ with d regular and

$$E = \{x \in R \mid ax \in dR\}.$$

Then dR is essential and hence so is E, so that E contains a regular element d_1. Thus, the right Ore condition is satisfied and R has a right quotient ring Q.

Next suppose that F is an essential right ideal in Q, then $F \cap R$ is essential in R. Now $F \cap R$ has a regular element, which is a unit in Q, and hence $F = Q$. Let J be a right ideal and K a right ideal of Q such that $J \cap K = 0$ and $J \oplus K$ is essential (use Zorn's lemma); then $J \oplus K = Q$. Thus, the module Q_Q is semisimple and Q is a semisimple ring.

Conversely, let R have a semisimple right quotient ring Q. Then a right ideal E of R is essential if and only if $EQ = Q$. To see this, suppose I is a nonzero right ideal of Q, then $I \cap R \neq 0$ and $I \cap R \cap E \neq 0$, taking E to be essential in R. Hence $I \cap EQ \neq 0$, which means that EQ is essential in Q; then $EQ = Q$ as Q_Q is a direct sum of simple modules. On the other hand, when $EQ = Q$ is given and I is a nonzero right ideal of R then $IQ \cap EQ \neq 0$, trivially, and hence $I \cap E \neq 0$, so that E is an essential right ideal of R. These conditions are equivalent to saying that E has a regular element, because whenever $1 \in EQ$, $1 = ec^{-1}$, $e \in E$, $c \in R$ and regular. On the other hand, when E has a regular element, then $EQ = Q$ and E is essential in R.

We now conclude that R is a semiprime ring, because if N is a nilpotent ideal of R, ${}^\perp N$ is essential as a right ideal and hence, has a regular element. Thus $N = 0$.

Let S be a direct sum of nonzero right ideals $\{I_\alpha \mid \alpha \in A\}$ of R which is essential as a right ideal. S has a regular element c, expressible as a finite sum

$$c = x_1 + \ldots + x_n; \, x_i \in I_{\alpha_i}.$$

Now cR is essential and lies in $I_{\alpha_1} + \ldots + I_{\alpha_n}$; it follows that A has only the indices $\alpha_1, \ldots, \alpha_n$, and R has (a.c.c.)$\oplus$.

Finally, the maximum condition holds for right annihilators, because R is a subring of the Noetherian ring Q. □

4.5 COROLLARY *A prime ring R has a simple Artinian right quotient ring if and only if R satisfies (a.c.c.)$\oplus$ and (a.c.c.)$^\perp$.* □

A set $M = \{e_{ij} \mid 1 \leq i, j \leq n\}$ of elements of R is called a set of $n \times n$ **matrix units** of R if

$$e_{ij}e_{kl} = \delta_{jk}e_{il}$$

and

$$\sum_{i=1}^{n} e_{ii} = 1$$

for all i, j, k and l. (δ_{jk} is the Kronecker delta.)

Faith–Utumi theorem

If F is a right order in a ring D, then the $n \times n$ matrix ring F_n is a right order in D_n, and so is any ring R such that $F_n \subseteq R \subseteq D_n$. The theorem below states that if D is a semisimple ring, every order R of D_n is obtained this way.

4.6 FAITH-UTUMI THEOREM [65] *Let R be any subring-1 of a matrix ring $S = D_n$, where D = centralizer $_SM$ of a set $M = \{e_{ij}\}_{i,j=1}^{n}$ of matrix units. Let*

$$A = A_M = \{a \in R \mid aM \subseteq R\}$$

and

$$B = B_M = \{b \in R \mid Mb \subseteq R\}.$$

Then

$$BA = (BA \cap D)_n = \sum_{i,j=1}^{n} (BA \cap D)e_{ij}$$

is a subring-1 of R, AB is an ideal of R, and $(BA)^2 \subseteq AB$. Furthermore, $F = BA \cap D$ and $U = B \cap D = A \cap D$ are *ideals of* $P = R \cap D$ *such that* $F \supseteq U^2$. *Thus,* $R \supseteq BA = F_n \supseteq U_n{}^2$.

If D_n *is a right quoring of* R, *then* M *can be so chosen so that* A *contains a unit* a *of* S, B *contains a unit* b *of* S, *and* $BA = F_n$ *is a right order of* D_n *equivalent to* R *and to* P_n. *If* R *is also a left order of* D_n, *then any choice of matrix units* M *has these properties.*

When D *is a semisimple*† *ring and* D_n *is a right quoring of* R, *then* D *is the right quoring of* F, *and*

$$D_n = \{af^{-1} \mid a \in F_n, \text{ regular } f \in F\}.$$

Proof Since $AP \subseteq A$, $PB \subseteq B$, and $P \subseteq D$, then $PF \subseteq F$, and $FP \subseteq F$, so that F is an ideal of P. Since B is a right ideal of R, then $U = B \cap D$ is a right ideal of $P = R \cap D$, and $U = A \cap D$ is a left ideal of P. Furthermore,

$$U \supseteq F = BA \cap D \supseteq (B \cap D)(A \cap D) = U^2.$$

Now, every element $t \in D_n$ has a unique representation $t = \sum_{i,j=1}^{n} t_{ij}e_{ij}$, where

$$t_{ij} = \sum_{k=1}^{n} e_{ik}te_{kj} \in D.$$

Thus, if $t = ba \in BA$, $b \in B$, $a \in A$, then $t_{ij} \in BA \cap D = F$ since $MB \subseteq B$ and $AM \subseteq A$. This proves that $BA = F_n$. Moreover, $(BA)^2 = BABA \subseteq AB$, since A is a left, and B is a right ideal.

Now assume that D_n is a right quoring of R. First, let M' be any set of $n \times n$ matrix units of D_n, with centralizer D'. Clearly, there is a common denominator a in R of the elements in M', that is, a regular element $a \in R$ such that $M'a \subseteq R$. Then,

$$M = a^{-1}M'a$$

is a set of matrix units of S, and

$$D = a^{-1}D'a = \text{centralizer }_S M$$

is a ring isomorphic to D'. Furthermore, a is a unit of S contained in

$$A = \{x \in R \mid xM \subseteq R\}.$$

† This holds more generally for any semilocal quoring D_n (Faith [71]). The original theorem of Faith–Utumi [65] was stated for a field D. (Some of 4.6 is taken from Faith [65].)

Moreover

$$B = \{y \in R \mid My \subseteq R\}$$

also contains a unit b of S. We now show that any set M of matrix units of S, for which A and B just defined each contains a unit of S, has the stated properties. Since every set M of matrix units satisfies this requirement if R is also a left order in S, then every set of matrix units has the stated properties in this case.

Since AB is an ideal of R, and since AB contains a regular element $t = ab \in R$, then D_n is a right quoring of AB, and every $y \in D_n$ has the form $y = xt(ct)^{-1}$, for elements x, regular $c \in R$. Since $bSb^{-1} = S$, where $S = D_n$, every element $y \in S$ has the form $y = by_1b^{-1}$, for some $y_1 = xr^{-1} \in S$, where x, regular $r \in R$. Thus,

$$y = b(xr^{-1})b^{-1} = (bxa)(bra)^{-1},$$

with $bxa \in BA$ and $bra \in BA$. This proves that D_n is a right quoring of BA. Now $bRa \subseteq BA$, and $BA \subseteq R$, so that BA is a right order of D_n equivalent to R. Moreover, since F is an ideal of P, then F_n is an ideal of P_n containing a regular element t. Thus, $tP_n \subseteq F_n$, so F_n is equivalent to P_n.

Assume that D_n is a semisimple right quoring of F_n. Since semiprime right Goldie rings are Morita invariant, F is semiprime right Goldie and, hence, F has a semisimple right quoring Q. (If F is a subring-1 of D, then the subring F' generated by F is semiprime right Goldie, with semisimple right quoring Q, and then Q is the right quoring of F.) Since every regular element $f \in F$ is regular in F_n (identifying f with diag f), $f^{-1} \in D_n$, and in fact, $f^{-1} \in D$.

Q can be embedded in D by a mapping which induces 1_F. Then, F_n is a right order in Q_n, and

$$Q_n = \{ac^{-1} \mid a \in F_n, c \in F\}.$$

However, Q_n is semisimple, and hence every right regular element of Q_n is a unit, so Q_n is a right quoring of F_n. This proves $Q_n = D_n$, so that $Q = D$ is the right quoring of F. □

Principal ideal rings

If R is a right order in D_n, the theorem of Faith and Utumi states that there exists an order F in D and an inclusion $F_n \subseteq R \subseteq D_n$. Even though F_n is an order in D_n, the ring structure of R does not depend entirely on F_n. Under a stronger hypothesis, however, it does.

4.7 PRINCIPAL RIGHT IDEAL THEOREM (Goldie [62]) *If R is a prime ring and a principal right ideal ring, then there exist a right Ore domain F, an integer $n > 0$, and an isomorphism $R \approx F_n$.*

Proof (Faith–Utumi [65]) Using the notation of 4.6, we can write $B = cR$, for some $c \in R$. Since B contains a regular element, c is a regular element. Since $e_{ij}B \subseteq B$, that is, $e_{ij}cR \subseteq cR$, then

$$c^{-1}e_{ij}c = f_{ij} \in F \quad (i, j = 1, \ldots, n).$$

Now $M = \{f_{ij} \mid i, j = 1, \ldots, n\}$ is a complete set of matrix units in S, with associated field $H = c^{-1}Dc$. Since $M \subseteq R$, $R = \sum_{i,j=1}^{n} f_{ij}K$, where $K = H \cap R$. Since R is an order in H_n, then K is an order in H; that is, K is an Ore domain. □

F need not be a principal ideal ring (cf. Swan [62]). Nevertheless, F is right Noetherian and hereditary, since these properties are Morita invariant.

Supplementary remarks (1) Let R be a right order in the ring $S = D_n$, $n > 1$. If f is any idempotent of S such that fS is a minimal right ideal, then $f = ac^{-1}$, with $a, c \in R$. Thus $e = c^{-1}fc$ is idempotent and $0 \neq ce \in Se \cap R$. Since $Se \cap R \neq 0$, it follows from primeness of R that $(eS \cap R)(Se \cap R) \neq 0$, so that $eSe \cap R \neq 0$. Then the proof of 4.6 shows that $F = eSe \cap R$ is a right Ore domain, and $D \approx eSe$ is its right quotient field; we thus obtain that $S \approx D_n$, where D contains a right Ore domain K, which is contained in R. This illustrates the precise nature of 4.6 which states much more.

(2) Next we show that, in general, not every set M of matrix units has the property without the hypothesis that R is also a left order in S.

Let K be a right Ore domain with right quotient field D that is not a left Ore domain. Let x, y be nonzero elements of K such that $Kx \cap Ky = 0$, and let

$$R = \begin{pmatrix} Kx & Ky \\ Kx & Ky \end{pmatrix}.$$

[R is the set of all 2×2 matrices $\begin{pmatrix} a & b \\ c & d \end{pmatrix}$ with $a, c \in Kx$, and $b, d \in Ky$.] Since K is right Ore, if $A = \begin{pmatrix} a & b \\ c & d \end{pmatrix}$ is an arbitrary element of D_2, there exists $q \in K$, $q \neq 0$, such that aq, bq, cq, $dq \in K$, and then

$$B = \begin{pmatrix} a & b \\ c & d \end{pmatrix} \begin{pmatrix} qx & 0 \\ 0 & qy \end{pmatrix} \in R.$$

Thus, $A = BC^{-1}$, with $B, C = \begin{pmatrix} qx & 0 \\ 0 & qy \end{pmatrix} \in R$. Hence, R is a right order in D_2.

As in the theorem, identify D with the subring of D_2 consisting of all scalar matrices $\begin{pmatrix} k & 0 \\ 0 & k \end{pmatrix}$ with $k \in D$. Now assume for the moment that R contains a subring F_2, where F is a domain $\subseteq D$. The contradiction is immediately evident since the form of R,

$$R = \begin{pmatrix} Kx & Ky \\ Kx & Ky \end{pmatrix},$$

where $Kx \cap Ky = 0$, precludes the possibility of its containing a nonzero scalar matrix $\begin{pmatrix} d & 0 \\ 0 & d \end{pmatrix}$ with $d \in K$. $\square$

(3) A ring R, as in 4.6, need not contain nontrivial idempotents. Perhaps the simplest example is as follows: Let $S = \mathbb{Q}_2$ be the ring of all 2×2 matrices over the rational number field $\mathbb{Q}$, and let R be the subring consisting of all matrices $\begin{pmatrix} a & b \\ c & d \end{pmatrix}$, where b, c are even integers, and a, d are integers which are either both even or both odd. Then $R = (2\mathbb{Z})_2 + \mathbb{Z}$ is not an

integral domain. However, R has classical quotient ring S but does not contain idempotents $\neq 0, 1$.

Maximal orders

Throughout this section, all rings will be at least right orders in semisimple rings, i.e., semiprime right Goldie rings. The term order will be reserved for a semiprime two-sided Goldie ring. Recall that a right order R is **maximal (left-equivalent, right-equivalent)** provided that $R \subseteq R' \subseteq Q$ and $R \sim R'$ ($R \overset{l}{\sim} R'$, $R \overset{r}{\sim} R'$) imply that $R = R'$.

4.8 PROPOSITION (*a*) *Let R, S be right orders with $R \subseteq S$, $R \sim S$. There exist right orders T, T' such that*

$$R \subseteq T \subseteq S, \quad R \overset{l}{\sim} T \overset{r}{\sim} S,$$
$$R \subseteq T' \subseteq S, \quad R \overset{r}{\sim} T' \overset{l}{\sim} S.$$

(*b*) *R is maximal iff R is a maximal left- and right-equivalent order.*

Proof Consult Faith [73*a*], p. 413–4. □

4.9 PROPOSITION *A simple Goldie ring R is a maximal order in S.*

Proof Suppose $S \supseteq R$ and $S\alpha \subseteq R$ for some regular element $\alpha \in R$. Clearly, $S\alpha R = R$ by simplicity of R. Since $S^2 = S$, $SR = S(S\alpha R) = S^2\alpha R = S\alpha R = R$. However, $SR = S$. Thus, $S = R$ and hence R is a maximal left-equivalent order. Similarly, R is a maximal right-equivalent order and hence, by 4.8, a maximal equivalent order in Q. □

The essential argument used here was first used by Faith [65] who showed, *interalia*, that center R coincides with center Q.

4.10 PROPOSITION *Let R be a maximal right order, M_R a finite dimensional torsionless generator and $k = \operatorname{End} M_R$. If $k' \supseteq k$ and $k' \overset{r}{\sim} k$, then $R = \operatorname{End}_{k'}k' \otimes {}_kM$.*

Proof Set $M' = k' \otimes {}_kM$ and $R' = \operatorname{End}_{k'}M'$. Since ${}_{k'}M'$ is finitely generated and projective, $M'_{R'}$ is a generator and hence, balanced, i.e., if $k'' = \operatorname{End} M'_{R'}$, $R' = \operatorname{End}_{k''}M'$. Moreover, if ${}_kM = k^ne$, ${}_{k'}M' = k'^ne$ and $R' = ek'_ne$ implying that R' is an order in the quotient ring of R. Thus, End $M'_{R'}$ = End M'_R. Since $\alpha k' \subseteq k$, where α is a regular element of k, α induces a monomorphism $M'_R \to M_R$ implying that M'_R is finite dimensional, torsionless and hence, T = trace $M'_R \neq 0$. Since M'_R is clearly faithful, $TR' \subseteq R$ by 1.6. However, T_R is essential since M'_R is faithful and torsionless. Thus, T contains a regular element of R implying that $R' \overset{r}{\sim} R$ and hence $R' = R$ by maximality of R. □

4.11 PROPOSITION† *Let R be a maximal order, M_R a finite dimensional torsionless generator, k = End M_R, M^* =* $\hom_R(M, R)$ *and* $M^\# = \hom_k(M, k)$. *Then*

(*a*) End $M^\#{}_k$ = End $M^*{}_k = R$

(*b*) k' = End ${}_R M^\#$ = End ${}_R M^*$ *is an essential extension of k and hence, an equivalent order in the quotient ring of k.*

(*c*) dim ${}_R M^*$ = dim M_R

(*d*) $M^* \approx M^{***}$.

Proof (*a*) Set R' = Biend ${}_R M^\#$ = End $M^\#{}_k$ and T = trace ${}_R M^\#$. By 1.6, $R'T \subseteq R$ and since ${}_R M^\#$ is a generator, $T = R$. Thus, $R' = R$.

(*b*) Since k' = Biend $M^\#{}_k$, $Tk' \subseteq k$ where T = trace $M^\#{}_k$. Hence, k' is an essential extension of k and, since T is an essential right ideal of k, T contains a regular element of k and thus, $k' \overset{r}{\sim} k$.

(*c*) By 1.18(1), dim M_R = dim k_k and dim ${}_R M^*$ = dim ${}_{k'}k'$. Since both k and k' are orders and ${}_k k'$ is an essential extension of ${}_k k$, dim k_k = dim ${}_k k'$ = dim ${}_{k'}k'$ as claimed.

(*d*) By (*c*), dim ${}_R M^*$ = dim $M^{**}{}_R$ = dim ${}_R M^{***}$. Since ${}_R M^*$ is isomorphic to a direct summand of ${}_R M^{***}$ (see Jans [64], p. 67), we have ${}_R M^* \approx {}_R M^{***}$. □

The next lemma allows us to assume in many instances that the module in question is a generator (Jategaonkar [71]).

4.12 LEMMA *Let R be a maximal right order, M a finite dimensional, torsionless, faithful R-module and k = End M_R. Then k is a maximal (right-equivalent, left-equivalent) order when-*

† More generally, 4.11 holds for finite dimensional, torsionless, faithful M_R. The proof is almost identical.

ever $\operatorname{End}(M \oplus R)_R$ *is a maximal (right-equivalent, left-equivalent) order.*

Proof Viewing the elements of $M \oplus R$ as column vectors,

$$\bar{k} = \operatorname{End}(M \oplus R)_R = \begin{pmatrix} k & M \\ M^* & R \end{pmatrix}$$

and its quotient ring $\bar{D}$ is

$$\bar{D} = \operatorname{End}((M \oplus R) \otimes {}_R Q)_Q = \begin{pmatrix} \operatorname{End}(M \otimes {}_R Q)_Q & M \otimes {}_R Q \\ \hom_Q(M \otimes {}_R Q, Q) & Q \end{pmatrix},$$

where $Q =$ the quotient ring of R. Furthermore, $D = \operatorname{End}(M \otimes {}_R Q)_Q =$ the quotient ring of k (see Zelmanowitz [67]).

Suppose $\bar{k}$ is maximal and $k' \supseteq k$ with $k' \sim k$, say, $\alpha k' \beta \subseteq k$ with α and β regular elements of k. Set

$$S = \begin{pmatrix} k' & k'M \\ M^*k' & M^*k'M \end{pmatrix} + \bar{k}.$$

S is clearly an order in $\bar{D}$ containing $\bar{k}$. Since

$$\bar{k}\begin{pmatrix} \alpha & 0 \\ 0 & 0 \end{pmatrix}\bar{k}S\bar{k}\begin{pmatrix} \beta & 0 \\ 0 & 0 \end{pmatrix}\bar{k} \subseteq \bar{k}$$

(routine computation), $S \sim \bar{k}$ if the ideals $\bar{k}\begin{pmatrix} \alpha & 0 \\ 0 & 0 \end{pmatrix}\bar{k}$ and $\bar{k}\begin{pmatrix} \beta & 0 \\ 0 & 0 \end{pmatrix}\bar{k}$ contain regular elements of $\bar{k}$. To demonstrate this, we shall consider only the first ideal. Since $\bar{k}\begin{pmatrix} \alpha & 0 \\ 0 & 0 \end{pmatrix}\bar{k}$ contains all matrices of the form

$$\begin{pmatrix} k\alpha k & 0 \\ 0 & [M^*, \alpha M] \end{pmatrix}$$

such a regular element exists provided the ideal $I = [M^*, \alpha M]$ is essential as a right ideal of R; equivalently, $Ir = 0 \Rightarrow r = 0$. If $Ir = 0$, $\alpha M r = 0$ since M is torsionless. Thus, $Mr = 0$ (α is regular) $\Rightarrow r = 0$ since M is faithful. Consequently, I is essential as asserted.

The proofs of the parenthetical statements are special cases of the above. □

When R is prime, Hart–Robson [70] have shown that eR_ne is maximal for all idempotents $e \in R_n$ and $n \geq 1$. The argument used in the proof of 4.12 is essentially theirs.

4.13 THEOREM *Let R be a maximal order and M a finite dimensional, torsionless, faithful R-module. Then* $\text{End}\,_R M^* = m$ *is a maximal order and is the maximal right equivalent order containing k.*

Proof It suffices to show that whenever $k' \supseteq k = \text{End}\, M_R$ and $k' \overset{r}{\sim} k$, $k' \subseteq m$ and if $k' \supseteq m$ with $k' \overset{l}{\sim} m$, $k' \subseteq m$. For if this condition is satisfied and $k' \supseteq m$ with $k' \sim m$, $k' \sim k$ and by 4.8 there exists an order s with $k' \supseteq s \supseteq k$, $k' \overset{l}{\sim} s$ and $s \overset{r}{\sim} k$. By assumption, $s \subseteq m$ and since $k' \overset{l}{\sim} s$, $k' \overset{l}{\sim} m$. Thus, $k' \subseteq m$, and m is maximal as claimed.

To show that the above condition is fulfilled, we can clearly assume that M is a generator by 4.12. If $k' \supseteq k$ with $k' \overset{r}{\sim} k$, set $M' = k' \otimes_k M$. By 4.10, $R = \text{End}\,_{k'} M'$ and clearly, $\hom_{k'}(M', k') \to \hom_R(M', R)$ is a monomorphism. Since M_R is a generator, $R = \text{End}\,_k M$ and $M^{\#} \approx M^*$. Since M'_R is an essential extension of M_R, $\hom_R(M', R) \hookrightarrow \hom_R(M, R)$. One readily checks that the composition of maps

$$\hom_{k'}(M', k') \to \hom_R(M', R) \hookrightarrow \hom_R(M, R) \to M^{\#}$$

is the map defined by

$$f \to f|_M \text{ (restriction to } M) \quad \forall f \in \hom_{k'}(M', k').$$

Thus, any k-linear map $f: M \to k'$ is actually a map *into* k since f induces a k'-linear $\bar{f}: M' \to k'$ such that $\bar{f}|_M = f$.

Next, if $T = \text{trace}\,_k M$, $Tk' \subseteq k$. For, if $(u)f \in T$, $\alpha' \in k'$, $f\alpha' \in \hom_k(M, k') = \hom_k(M, k)$ by the above remarks. Hence, $(u)f\alpha' \in k$. Finally, $M^{\#}Tk' = M^{\#}k' \subseteq M^{\#}$ which implies that $k' \subseteq m$.

In order to complete the proof, suppose that $k' \supseteq m$ with $k' \overset{l}{\sim} m$. Then by an obvious modification of 4.10, $R = \text{End}(M^{\#} \otimes_m k')_{k'}$. By modifying the above argument, we obtain $\hom_m(M^{\#}, k') = \hom_m(M^{\#}, m)$ and hence, $k'T^{\#} \subseteq m$ where

$T^\# = \text{trace } M^\#{}_m$. Since $m = \text{End } M^{**}{}_R$ (see the proof of (*a*) $\Rightarrow$ (*c*) in 4.14) and $T^\# M^{**} = M^{**}$, $k' \subseteq m$ as claimed. □

4.14 MAXIMAL ORDER THEOREM *Let R be a maximal order and M a finite dimensional, torsionless, faithful R-module. Then* (*a*) $\Rightarrow$ (*b*) $\Leftrightarrow$ (*c*), *where*

(*a*) M_R *is R-reflexive,*

(*b*) $k = \text{End } M_R$ *is a maximal order,*

(*c*) $k = \text{End }{}_R M^*$, *where*

$$M^* = \text{hom}_R(M, R).$$

Moreover, if M is a generator, then (*b*) $\Rightarrow$ (*a*).

Proof We shall first assume that M is a generator and prove the stated equivalences for this case. Then we shall discuss how the general case can be reduced to the case of a generator.

(*a*) $\Rightarrow$ (*c*): Since $M = M^{**}$, $k = \text{End } M_R = \text{hom}_R(M, M^{**})$

$$= \text{hom}_R(M^*, \text{hom}_R(M, R)) = \text{End }{}_R M^*.$$

The third equality follows from one of the standard functorial isomorphisms (e.g., see Cartan–Eilenberg [56], ex. 4, p. 32).

(*c*) $\Leftrightarrow$ (*b*): Immediate by 4.11 and 4.13.

(*b*) $\Rightarrow$ (*a*): By 4.11, $k' = \text{End }{}_R M^\# = k$. Since $M^\# \approx M^*$ and

$$\text{hom}_{k'}(M^\#, k') \approx \text{hom}_R(M^*, R) = M^{**},$$

$M^{\#\#} \approx M^{**}$ as (k, R)-bimodules. By finite projectivity of ${}_k M$, $M^{\#\#} \approx M$ as (k, R)-bimodules. Hence, M is R-reflexive as asserted.

If M is no longer assumed to be a generator, note that M is reflexive iff $\bar{M} = M \oplus R$ is reflexive, and End $\bar{M}_R$ is maximal only if End M_R is maximal. Moreover, $\bar{M}$ is always a generator for arbitrary M. Thus, (*a*) $\Rightarrow$ (*b*), and (*b*) $\Rightarrow$ (*c*) follows immediately from 4.11(*b*). If $k = \text{End }{}_R M^*$, k is maximal by 4.13. Consequently, (*c*) $\Rightarrow$ (*b*). □

Remark To see that (*b*) $\Rightarrow$ (*a*) does not hold in general, let R be any maximal order having a faithful ideal I which is not reflexive as a right ideal and set $k = \text{End } I_R$; e.g., any n-dimensional regular local ring with $n \geq 2$. These are maximal orders by Auslander–Goldman [60] and must have nonreflexive right ideals by Matlis [68]. Since $k \supseteq R$ and $k \sim R$, $k = R$. Thus, I_R is not reflexive but k is maximal.

In order to generalize the above to maximal right orders, care must be exercised to insure that all of the endomorphism rings in question are right orders in the appropriate semisimple ring. For example, if M_R is a finite dimensional torsionless generator, there is no guarantee that $k' = \text{End}\,_R M^*$ is a right order in the quotient ring of k. However, when R is prime and M_R reflexive, we can conclude that k is maximal. Specifically,

4.15 PROPOSITION *Let R be a prime maximal right order, M_R a finite dimensional reflexive R-module and $k = \text{End}\, M_R$. Then k is a maximal right order.*

Proof Since 4.12 is valid for right orders, it suffices to assume that M_R is a generator. Moreover, the proof of 4.13 shows that if $k' \overset{r}{\sim} k$ or $k' \overset{l}{\sim} k$, $Tk' \subseteq k$ provided that $R = \text{End}\,_{k'}k' \otimes_k M$. This, of course, implies that k is maximal. Thus, it suffices to show that if $k' \overset{l}{\sim} k$, $R = \text{End}\,_{k'}k' \otimes_k M$ (4.10 covers the case $k' \overset{r}{\sim} k$).

To that end note that for some $n > 0$, $R' = \text{End}\,_{k'}k' \otimes_k M \approx ek'_n e$ and $R \approx ek_n e$. Since $k'\alpha \subseteq k$, $ek'_n ek_n ek_n \bar{\alpha} ek_n e \subseteq ek_n e$ where $\bar{\alpha} = \alpha I_n$, I_n the $n \times n$ identity matrix. Since $ek_n e$ is prime, the ideal $ek_n ek_n \bar{\alpha} ek_n e$ is essential and hence, contains a regular element of $ek_n e$. Thus, $ek'_n e \overset{r}{\sim} ek_n e \Rightarrow ek'_n e = ek_n e$ by maximality of $R(ek_n e)$. □

The question of whether one-sided versions of 4.13 and 4.14 hold for general semiprime R remains open. If $k' \overset{l}{\sim} k$ always implies $ek'_n e \overset{l}{\sim} ek_n e$, then, of course, the above argument shows that finite dimensional, faithful, reflexive modules have maximal endomorphism rings.

As an interesting aside, note that the proof of 4.15 shows that $m = k_T$, the (left) quotient ring of k with respect to the smallest filter of all left ideals which contain $T = \text{trace}\,_k M$.

The case of global dimension at most 2

In order to show that reflexive generators, specifically in the case of maximal orders, are bonafide generalizations of finitely generated projective generators, we shall show (4.17) that if each finitely generated reflexive module is projective, then the global dimension of the ring cannot exceed 2 (the converse is well-known). Then, if U_R is any finitely generated reflexive nonprojective (necessarily r.gl.dim $R > 2$), $U_R \oplus R_R$ is a non-projective reflexive, generator. For simplicity, we shall assume that R is a semiprime two-sided Noetherian ring. To prove the above assertion we need the following characterization of two-sided Noetherian rings of global dimension ≤ 2.

4.16 PROPOSITION (Bass [60]) *If R is left and right Noetherian, the following are equivalent:*

(*a*) r.gl.dim $R \leq 2$.

(*b*) *The dual of any finitely generated right R-module is projective* □

Let M_R be any finitely generated R-module and $t(M)$ its torsion submodule. We can assume that M_R is a generator by the remarks preceding the conclusion of the proof of 4.14. Clearly, $M^* = (M/t(M))^*$ since $(t(M))^* = 0$. Thus, we can further assume that M is a torsion free and hence, a torsionless, finitely generated R-module. Since M_R is clearly finite dimensional, ${}_RM^* \approx {}_RM^{***}$ and assuming reflexive modules are projective, M^* is projective. Consequently, by 4.16, r.gl.dim $R \leq 2$. Thus, we have shown that

4.17 PROPOSITION *The following conditions are equivalent for any semiprime two-sided Noetherian ring R.*

(*a*) r.gl.dim $R \leq 2$.

(*b*) *each finitely generated reflexive module is projective.* □

A canonical and important example of a reflexive module is given by

4.18 PROPOSITION *Any maximal uniform right ideal of R (more generally, any right annulet) is reflexive.*

Proof By Bass [60] $U = X^*$ for some finitely generated left R-module X. Since we can assume that ${X^*}_R$ is a generator, $\dim {X^*}_R = \dim {X^{***}}_R$ and consequently, $U = X^* = X^{***} = U^{**}$. □

By parroting an argument used in Proposition 2.26 we readily obtain the necessity of the following characterization of simple Noetherian domains having global dimension less than or equal to 2.

4.19 THEOREM (*a*) *Let R be a simple two-sided Noetherian domain. Then* r.gl.dim $R \leq 2$ *iff* $\forall n \geq 1$, *whenever B is a maximal order and* $B \overset{Q}{\sim} R_n$, *B is simple.* (*b*) *If R is simple Noetherian with* r.gl.dim $R \leq 2$, *M_R finite dimensional and torsionless, then* $k' =$ End ${}_RM^*$ *is a simple Noetherian ring Morita equivalent to R.*

Proof (*b*) Since M_R is finite dimensional and torsionless, $M_R \hookrightarrow R^{(n)}$ for some $n \geq 1$, implying that T = trace $M_R \neq 0$. By simplicity of R, $T = R$ and hence M_R is a generator. Therefore, by 4.16 ${}_RM^*$ is a progenerator and hence, k' is Morita equivalent to R.

(*a*) (Necessity): We shall present only the $n = 1$ case. The general case is identical. Suppose $xBy \subseteq R$. Set $a = xy$ and $B' = xBx^{-1}$. Then $B'a \subseteq R$ and hence $I = B'aR$ is a right ideal of R satisfying $B' \subseteq$ End I_R. Since End $I_R \sim R$, $B' =$ End I_R by maximality of B' (B' maximal iff B is). Moreover, $B' =$ End ${}_RI^*$ by 4.11. Since r.gl.dim $R \leq 2$, ${}_RI^*$ is a progenerator implying that B' and hence, B is simple.

(Sufficiency): By 4.17 it suffices to show that each finitely generated, reflexive R-module is projective. To that end suppose I_R is finitely generated reflexive and dim $I_R = n$. Then clearly $R^{(n)} \hookrightarrow I_R$ and since dim ${}_RI^* = n$, $I_R \hookrightarrow R^{(n)}$. This readily implies that $B =$ End $I_R \sim R_n =$ End ${R_R}^{(n)}$. B is a maximal order since I_R is a reflexive generator and hence, simple by hypothesis. Finally, by simplicity of B, I_R is projective, completing the proof of 4.19. □

As a final remark regarding uniform right ideals we have

4.20 REFLEXIVE IDEAL THEOREM *If R is a simple two-sided Noetherian ring satisfying* r.gl.dim $R \leq 2$ *and each uniform right ideal is reflexive, then R is hereditary.*

Proof We can assume R is a domain by 4.19. Since each right ideal is now reflexive, each right ideal is projective by 4.17.

Thus, even in such well-behaved rings, one cannot expect to find too many reflexive uniform right ideals.

The nonreflexive case

The object of this section is to provide further insight into the structure of the endomorphism ring of a finite dimensional torsionless R-module M, in particular, a basic right ideal of R, and to sharpen Faith's representation theorem mentioned in the introduction. In 4.21 we shall assume that R is simple. While 4.21(b) is 4.13 in a less general setting, its proof is an interesting application of the correspondence theorem.

4.21 PROPOSITION *Let R be a simple order, M_R a finite dimensional torsionless R-module and $k = \text{End}\, M_R$. Then*

(a) *k is a maximal left equivalent order:*

(b) *$m = \text{End}\, {}_R M^*$ is the maximal equivalent order containing k.*

Proof (a) M_R is necessarily a generator for mod-R. Let $T = \text{trace}\, {}_k M$. If $k' \supseteq k$ with $k' \overset{l}{\sim} k$, say $k'\alpha \subseteq k$, then $k'\alpha T = T = k'T$ since T is the least ideal of k. Thus, $k'TM = k'M \subseteq M$ and hence $k' \subseteq k$.

(b) If $k' \supseteq k$ with $\alpha k' \subseteq k$, $T\alpha k' = T = Tk' \Rightarrow k' \subseteq m$ since $M^*T = M^*$. Thus, *any* equivalent order containing k is contained in m implying that m is the maximal such order. □

4.22 PROPOSITION (a) (Faith [64]) *Any maximal order in $Q = D_n$, D a field, is isomorphic to the biendomorphism ring of any right ideal U, say $R \approx \text{End}\, {}_k U$ where $k = \text{End}\, U_R$.*

(b) *If U_R is a maximal uniform right ideal, then $k = \text{End}\, U_R$ is a maximal order in D.*

Proof (a) Let U be any right ideal of R. By 1.6, $T = \text{trace}\, U_R$ satisfies $T\bar{R} \subset R$ where $\bar{R} = \text{Biend}\, U_R$. Since T contains a regular element of R, $\bar{R}$ is a right order of Q which is $\overset{r}{\sim}$ to R. Thus, $\bar{R} = R$ by maximality of R.

(b) Clear. □

Examples

In this section we show that if $A = \text{End}\, {}_B V$ is simple with ${}_B V$ finitely generated projective and B right Ore, B can be quite 'bad' when A is quite 'good'.

In the following examples, we shall freely use results from Robson [72] and certain generalizations which can be found in Goodearl [73]. These references will be abbreviated to [72] and [73] respectively.

EXAMPLE 1 For any $n \geq 2$, there exists a right but not left Noetherian domain B with l.gl.dim B = r.gl.dim $B = n$, a finitely generated projective left B-module ${}_BV$ such that A = End ${}_BV$ is a simple pri(pli)-domain.

Let $k \supset k'$ be fields with l.dim ${}_{k'}k = \infty$ and r.dim $k_{k'}$ finite (see Cohn [61*a*] for such examples). Set $A = \mathscr{D}_{k(x)}$ where δ is defined in the obvious fashion:

$$\delta(x) = 1,$$
$$\delta(a) = 0 \quad \forall a \in k.$$

By Example 7.3 in [72], the subring $B = k + \delta A$ of A is a hereditary Noetherian domain. By Theorem 4.3 of [73], $B' = k' + \delta A$ is a two-sided order in A with l.gl.dim B' = r.gl.dim B' = 2. Moreover, B' is right Noetherian and not left Noetherian.

Finally, set $V = \delta A$. First, V is finitely generated and projective in B'-mod. For $V^*{}_{B'} \supseteq A$ and $AV = A$. Thus, $1 \in V^*V$ implying finite projectivity of ${}_{B'}V$ by the dual basis lemma. Moreover, End ${}_{B'}V$ is an equivalent order containing A and hence, A = End ${}_{B'}V$.

By an obvious modification of this technique, we can choose B' such that B' is a two-sided order in A, B' is right but not left Noetherian and

$$\text{l.gl.dim } B' = \text{r.gl.dim } B' = n.$$

Moreover, B' can be chosen to be neither right nor left Noetherian but still a two-sided order, having any prescribed global dimension ≥ 2.

EXAMPLE 2 (Real Pathology) Let R be a pri-domain, D its right quotient field and suppose that D/R is semi simple in mod-R. Then there exists a right hereditary, right Noetherian domain B, which is, however, a *two-sided order*, a finitely generated projective ${}_BV$ such that A = End ${}_BV$ is a simple pri(pli)-domain.

Set $S = \mathscr{D}_{D(x)}$ where δ is the standard formal derivative of $D(x)$, $T = D + \delta S$ and $k = R + \delta S$. As before, T is a hereditary Noetherian domain. Also $k \overset{Q}{\sim} S$ since $\delta S \subset k$. Thus, k is a two-sided order in F = right quotient field of S.

We claim that k is right hereditary. First, one readily verifies that the exact sequence of k-modules

$$0 \to T/\delta S \to S/\delta S \xrightarrow{\pi} S/T \to 0$$

is split exact (the obvious map is the desired right inverse to π) and hence, $pd(S/T)_k \leq pd(S/\delta S)_k$. Since S_k is projective by Lemma 2.1 in [72], $pd(S/\delta S)_k \leq 1$ and hence, $pd(S/T)_k \leq 1$, implying that T_k is projective. Since k is an order in T, it is easy to see that $T \otimes {}_kT = T$. As an immediate consequence, we have $pd\ M_k = pd\ M_T$ for all $M \in$ mod-T by a proof similar to that of Lemma 2.8 in [72]. Moreover, since D/R is semisimple in mod-R and isomorphic to T/k, T/k is semisimple in mod-R and hence, in mod-k.

Finally, let I be any right ideal of k. Then IT is projective in mod-k and IT/I is a direct summand of a coproduct of copies of T/k. Since T_k is projective, pd $T/k \leq 1$ implying that pd $IT/I \leq 1$. Hence, pd $I_k = 0$. Thus, k is right hereditary and since it is an order in a field, k is right Noetherian.

Note that if we set $V = \delta S, S = \text{End}\,_kV$. We also remark that such rings as R exist. For example the rings discussed in Chapter 5 work. The existence of an example principal on the right but not on the left remains an open problem. However, we feel very strongly that such examples do exist. Note that R in Example 2 does not have to be left Ore! Finally, if such an R exists (which is not left Noetherian), then k affords us with an example of a two-sided order which is right hereditary but not left hereditary, thus answering a question raised in Camillo–Cozzens [73].

5. *V-rings*

In 1956, Kaplansky showed that a commutative ring A is Von Neumann regular iff each simple A-module is injective. Thus, a very natural and important class of commutative rings could be completely characterized by their simple modules and a fundamental homological property, namely, injectivity. However, this property does not characterize noncommutative regular rings, since, as Faith showed, some regular rings have injective simple modules, while others do not, whereas Cozzens [70] constructed simple pli-domains, not fields having simple modules injective. We present these examples in this chapter. We also show that any right Goldie V-ring is a finite product of simple V-rings.

Dual to the notion of projectivity and of interest to us in this and succeeding chapters will be the concept of injectivity. For a proof of 5.1 and the statements regarding the injective hull, consult Faith [73*a*] or Lambek [66].

5.1 PROPOSITION AND DEFINITION *A module M_R is called* **injective** *if it satisfies any of the following equivalent conditions:*

(1) *The functor* $\hom_R(-, M) : \text{mod-}R \rightsquigarrow Ab$ *is exact.*

(2) *Every exact sequence*

$$0 \to M \to M' \to M'' \to 0$$

splits.

(3) (Baer's criterion) *If $f : I \to M$ is a map of a right ideal I of R into M, there exists an $m \in M$ such that*

$$f(x) = mx \quad \textit{for all} \quad x \in I. \qquad \square$$

Given any module M, there exists an injective module $\hat{M}_R$, unique up to isomorphism, satisfying the following equivalent properties:

(1) $\hat{M}_R$ is the maximal essential extension of M.

(2) $\hat{M}_R$ is the minimal injective extension of M.

$\hat{M}_R$ is called the **injective-hull** of M and is denoted variously by $\hat{M}$ or inj hull M_R.

There is an interesting connection between the ascending chain condition and injectivity, namely,

5.2 PROPOSITION *The following are equivalent for any ring R:*

(1) *R is right Noetherian.*

(2) *If $\{S_i : i \in I\}$ is any family of simple modules, $\sum_{i \in I} \oplus \hat{S}_i$ is injective.*

(3) *Each direct sum of injective modules is injective.*

Proof (After Bass) We shall show that (2) $\Rightarrow$ (1) leaving the other implications for the reader.

Let $I_1 \subseteq I_2 \subseteq \ldots$ be an ascending chain of finitely generated right ideals. Since the I_j are finitely generated, we can embed the above chain in

$$I_1 \subseteq M_2 \subset I_2 \subseteq M_3 \subset I_3 \ldots,$$

where I_j/M_j is simple $\forall j \geq 2$. Set $E_j = \widehat{I_j/M_j}$ and $E = \sum_{j \geq 2} \oplus E_j$. By hypothesis, E is injective. If $I = \cup I_j$, there is a map $f : I \to E$ defined as follows:

$$f(i) = \sum_{j \geq 2} i + M_j \quad \forall i \in I.$$

The sum on the right is of course finite since eventually, $i \in I_j$, $\forall j \geq n$. By 5.1, f extends to all of $R \Rightarrow \operatorname{im} f \subseteq$ finite summand of E, say $\sum_{j=2}^{n} \oplus E_j$. This readily implies that $I_{n+1} = I_{n+j}$, $\forall j \geq 2$ and hence the chain terminates. □

There is another useful characterization of Noetherian rings, due to Matlis [58] and Papp [59] (generalized by Faith and Walker [67]) which states that R is right Noetherian if and only if each injective module is a direct sum of indecomposable injectives. We remark that the necessity is an immediate consequence of 5.2, and the obvious fact that over a Noetherian ring, every nonzero module has a uniform submodule. The sufficiency is however, more difficult.

By modifying 5.1(1) we obtain the more general but very useful concept of quasi-injectivity.

5.3 DEFINITION $M \in$ mod-R is called **quasi-injective** if

$$0 \to N \to M \to M/N \to 0$$

exact implies

$$0 \to \hom_R(M/N, M) \to \hom_R(M, M) \to \hom_R(N, M) \to 0$$

exact, i.e., every map $f: N \to M$ where $N \subseteq M$ is induced by an endomorphism of M. Note that every semisimple module is quasi-injective, thus, attesting to their importance. There is a very easy way to identify these modules: they are the fully invariant submodules of their injective hull.

5.4 PROPOSITION (Johnson–Wong [61]) *M_R is quasi-injective if and only if $\Lambda M = M$ where $\Lambda = \operatorname{End} \hat{M}_R$.*

Proof That ΛM is quasi-injective is obvious. Conversely, suppose that M_R is quasi-injective and that $\lambda \in \Lambda$. We must show that $\lambda M \subseteq M$. Set $N = \{m \in M \mid \lambda m \in M\}$. If $N \neq M$, by quasi-injectivity, there exists a $\lambda_1 \in \operatorname{End} M_R$ such that $\lambda_1 m = \lambda m$, $\forall m \in N$. By injectivity of $\hat{M}_R$, there exists a $\lambda_1' \in \Lambda$ such that $\lambda_1' m = \lambda_1 m$, $\forall m \in N$. Since $\lambda_1 M \subseteq M$, $\lambda M \subseteq M$ if $(\lambda_1' - \lambda)M = 0$. Now $(\lambda_1' - \lambda)M \neq 0 \Rightarrow (\lambda_1' - \lambda)M \cap M \neq 0$ since $\hat{M} \bigtriangledown M$. However, if $0 \neq m' = (\lambda_1' - \lambda)m \in (\lambda_1' - \lambda)M \cap M$, $\lambda m = \lambda_1 m - m' \in M \Rightarrow m \in N \Rightarrow \lambda_1 m = \lambda m$ a contradiction. Thus, $N = M$ and the proof is complete. □

Although it will be used only briefly, we shall introduce the concept of a cogenerator which is dual to the notion of a generator.

5.5 PROPOSITION AND DEFINITION *A module C_R is called a* **cogenerator** *if it satisfies any of the following equivalent properties:*

(1) *The functor*

$$\hom_R(-, C) : \text{mod-}R \rightsquigarrow Ab$$

is faithful, i.e., given a nonzero $f: M \to N$ *there exists* $g: N \to C$ *such that* $g \circ f \neq 0$.

(2) *For each* $M \in$ mod-R *there exist a set* I *and an embedding* $M \to C^I$ *(product of* I *copies of* C*).*

(3) *For each* $M \in$ mod-R *the canonical map* $M \to C^{\hom_R(M,C)}$ *is a monomorphism.*

(4) *Given any simple module* S, *there is a monomorphism* $\hat{S} \to C$.

Proof The proofs of (1) $\Leftrightarrow$ (2) $\Leftrightarrow$ (3) are dual to those given for the dual statements in 1.1 and (1) $\Rightarrow$ (4) is obvious.

(4) $\Rightarrow$ (1) Let $f: M \to N$ be any nonzero map and $f(x) = y \in N$ be nonzero. $yR \approx R/y^\perp$ and $y^\perp \subseteq J$, a maximal right ideal.

By assumption, there exists a monic $h: \widehat{R/J} \to C$. Thus, we obtain a map $g': yR \to \widehat{R/J}$ defined by the composition

$$yR \to R/y^\perp \to R/J \to \widehat{R/J}.$$

By injectivity of $\widehat{R/J}$, g' extends to a map $g'': N \to \widehat{R/J}$. It is easy to check that the composition $g = h \circ g'': N \to C$ is the required map satisfying $g \circ f \neq 0$. $\square$

5.6 Definition A ring A is called a **right** V**-ring** if each simple right A-module is injective.

5.7 Theorem *Let* R *be a right* V*-ring. Then the following conditions are equivalent:*

(1) *Each simple right* R*-module is injective.*

(2) *Each right ideal is the intersection of maximal right ideals.*

(3) *The radical of* M, Rad $M = 0$ *for all* $M \in$ mod-R.

Proof (1) $\Rightarrow$ (3): Let S_i, $i \in I$ be a set of representatives of the distinct isomorphism classes of simple R-modules. Then, clearly, $C = \prod_{i \in I} S_i$ is an injective cogenerator in mod-R. Hence, for $M \in$ mod-R, there exists a monomorphism $0 \to M \to C^I$ (product of C, I times). This readily implies that rad $M = 0$.

(3) $\Rightarrow$ (2): Obvious.

(2) $\Rightarrow$ (1): Let f be a map of a right ideal I into R/M where M is an arbitrary maximal right ideal. To show that R/M is

injective, it suffices to show that f can be extended to R. $\operatorname{Ker} f = I_1$ is such that $xR + I_1 = I$ for some $x \in I$. By (2), there exists a maximal right ideal M_1 of R such that $I_1 \subseteq M_1$ and $M_1 + I = R$. Now

$$M_1 \cap I = M_1 \cap (xR + I_1) = I_1 + (xR \cap M_1) = I_1.$$

Therefore, by defining $\bar{f}$ to be zero on M_1 and $\bar{f} = f$ on I, we obtain the desired extension of f to R. □

Remark (1) $\Leftrightarrow$ (2) is credited to Villamayor (unpublished).

As an obvious corollary to the above argument, we have

5.8 COROLLARY *If R is a right V-ring, and I is any right ideal of R, then $I^2 = I$.* □

Note that this property is also enjoyed by any simple ring R. For $I = IR = IRIR = I^2$. Finally, we obtain

5.9 THEOREM *Let R be commutative. Then R is Von Neumann regular iff R is a V-ring.*

Proof If R is regular, so is R/I for any ideal I. Since $\operatorname{rad}(R/I) = 0$, condition (2) of 5.7 is trivially satisfied.

Conversely, if R is a V-ring and $a \in R$, then $aRaR = aR$ by 5.8. Thus $a^2x = a$ for some $x \in R$ and hence R is regular. □

In order to present Faith's example of a regular ring which is not a left V-ring (it is however a right V-ring!), we first summarize several useful categorial results.

5.10 PROPOSITION *If $T : \mathscr{C} \rightsquigarrow \mathscr{D}$ is a functor of abelian categories with left adjoint $S : \mathscr{D} \rightsquigarrow \mathscr{C}$ and if $\mathscr{C}$ has enough injectives, then T preserves injectives iff S is exact.*

Proof See Faith [73*a*], pp. 318 and 440. □

5.11 COROLLARY *The following properties are equivalent:*

(1) *Every injective A/B-module M is injective as a canonical right A-module.*

(2) *The inclusion functor* mod-$A/B \rightsquigarrow$ mod-A *preserves injectives.*

(3) *The inclusion functor* mod-$A/B \rightsquigarrow$ mod-A *preserves injective hulls of simple right A/B-modules.*

(4) *A/B is flat as a left A-module.* □

5.12 COROLLARY *If A is a right V-ring, then, for any ideal B, A/B is a flat left A-module.* □

5.13 COROLLARY *If A is a regular ring, then every injective right A/B-module is canonically an injective right A-module, for any ideal B of A.* □

5.14 EXAMPLE Let $L = \text{End } V_F$ be the full right linear ring of an infinite dimensional right vector space V over a field F, let S be the ideal consisting of linear transformations of finite rank, and let $R = S + F$ be the subring generated by S and the subring F consisting of scalar transformations (sending every $v \mapsto va$ for some $a \in F$). Clearly, $R(L)$ is a regular ring. By standard functorial arguments (${}_LV$ is flat, V_F is injective and $L = \text{End } V_F$), L_L is injective, and since V_L is a summand, V_L is likewise injective. Moreover, ${}_LV$ is injective iff $\dim V_F < \infty$ (see Faith [72*b*], p. 163).

First we show that R is a right V-ring. Let W be a simple right (or left) ideal of R, and let $f: I \to W$ be a mapping of a right (or left) ideal of R. In order to determine if W is injective, it suffices to assume that I is an essential one-sided ideal, and thus, that I contains the socle S of R (the sum of the simple right ideals of R). Since R/S is simple, either $I = S$, or $I = R$. If $I = S$, and f is a morphism of mod-R, then f is a morphism of mod-L, since

$$[f(xa) - f(x)a]s = 0 \quad \forall s \in S, x \in I, a \in L.$$

Since W is a summand of L in mod-L, then W is injective along with L, and so f has an extension to a mapping $L \to W$ which induces a mapping $R \to W$ extending f. This proves that W is injective in mod-R. If W is a simple right R-module not contained in S, then $W \approx R/S$ is injective over R/S, whence by 5.13 W is injective over R. This proves that R is a right V-ring.

In order to prove that R is not a left V-ring, assume for the moment that V is injective in R-mod, and that $f: I \to V$ is a morphism in L-mod, where I is a left ideal of L. In order to extend f, we may assume that I is an essential left ideal, that is, that $I \supseteq S$. Then, f has an extension f' in R-mod, and f' is a morphism in L-mod, since

$$s[(ax)f' - a(x)f'] = 0 \quad \forall s \in S, x \in I, a \in L.$$

But this implies that V is injective in L-mod, a contradiction. Thus, R is not a left V-ring. (This fact is also a consequence of Faith [67*a*], p. 103, Th. 3.1, which implies that the maximal left quotient ring of R is also a right quotient ring. This involves a contradiction.)

Next we prove a lemma preparatory to the structure theorem for a right Goldie V-ring.

5.15 LEMMA *If A is a right V-ring and I is an ideal of A, then I does not contain a right regular element of A.*

Proof Let M be a maximal right ideal containing I. Then A/M is injective and hence,

$$A/Mx = A/M \quad \forall \text{ right regular } x.$$

For, if $x^{\perp} = 0$ is arbitrary, there is a map $xA \to V$ defined by $xa \mapsto va$, $\forall a \in A$. By injectivity, $\exists v' \in V$ such that $v = v'x \Rightarrow V \subseteq Vx$. Finally, since $A/Mx = 0$, $\forall x \in I$, I cannot contain a right regular element of A. □

5.16 THEOREM *Let A be a right Goldie V-ring. Then A is a finite product of simple Goldie V-rings.*

Proof Since A is a right V-ring, rad $A = 0$, i.e., A is semiprime. Let I be a maximal ideal of A. If $I = 0$, A is already simple. Otherwise, $I \cap K = 0$ for some $K \neq 0$ by the lemma (as seen in the proof of 4.4, essential right ideals contain regular elements), and since A is semiprime, $I^{\perp} \neq 0$. Consequently, $A = I \oplus I^{\perp}$ and A/I is clearly a simple Goldie V-ring. The proof is now completed by induction on I, a Goldie V-ring having dimension less than dim A. □

It is instructive to give an alternate proof of this theorem using another useful result.

5.17 PROPOSITION *Let A be a prime Goldie V-ring. Then A is simple.* □

A ring R is the **subdirect product** of rings R/M_i, $i \in I$, if $\bigcap_{i \in I} M_i = 0$.

Alternate Proof If R is semiprime Goldie, the canonical ring homomorphism $R \to \prod_{i=1}^{n} R/M_i$ is an embedding, where R/M_i is a prime Goldie ring (see Jacobson [64*a*], p. 269). If R is a V-ring, so is each R/M_i and hence, each R/M_i is simple by the previous result. However, any subdirect product of finitely many simple rings is a direct product of simple rings. □

5.18 PROPOSITION *If A is a V-domain, then A is simple.*

Proof Let I be a nonzero ideal of A and $0 \neq a \in I$. $aAaA = aA$ by 5.8. Thus, $1 \in AaA$ which implies that $I = A$. □

(For the background of 5.16–18, see Ornstein [67] and Faith [67*a*, 72*a*, *b*].)

Let A be a hereditary 2-sided Noetherian right V-ring. By 2.33 each finitely generated torsion A-module (either side) is semisimple, and, since A is a right V-ring, each finitely generated torsion right A-module is injective. If in addition, A is a domain, we have,

5.19 PROPOSITION *Let A be a hereditary, two-sided Noetherian right V-domain. Then*

(1) *each cyclic right A-module not isomorphic to A is injective;*

(2) *each indecomposable injective right A-module is either simple or isomorphic to $\hat{A}$;*

(3) *each quasi-injective right A-module is injective.*

Proof (1) Clear.

(2) Let E be an indecomposable injective. Then $E = \hat{C}$ for any cyclic submodule C. If $C \approx A$, then $E \approx \hat{A}$. Otherwise, C is injective and $E = C$ which implies, of course, that E is simple.

(3) Let Q be quasi-injective. Then $\hat{Q} = \sum_{i \in I} \oplus E_i$, where each E_i is indecomposable and injective, and $E_i \cap Q \neq 0$, $\forall i \in I$.

Thus, we can assume that Q is indecomposable because $Q = \sum_{i \in I} \oplus (E_i \cap Q)$ and the obvious fact that each summand of a quasi-injective module is quasi-injective. By part 2, either $\hat{Q}$ is simple which implies that $Q = \hat{Q}$ or $\hat{Q} = \hat{A}$. Since Q is a left $\hat{A}$ submodule of $\hat{Q}$, then $Q = \hat{A}$ since $\hat{A}$ is a field. □

If A is not a domain, $A \sim K$ where K is a right Ore domain since A is hereditary. Clearly, K is also a right V-ring satisfying (3) of 5.19. Moreover, since each finitely generated torsion left K-module is semisimple, (2) of 5.7 is trivially satisfied. Thus, K is also a left V-ring implying that A is likewise.

We next consider a fundamental problem in the theory of formal differential (or difference) equations namely the problem of trying to determine when an arbitrary consistent system of homogeneous linear differential equations has a solution in an arbitrary differential ring A. As we shall see, a necessary and sufficient condition for the solvability of any consistent system is that ${}_{\mathscr{D}_A}A$ be an injective module. In particular, when $A = k$ is a field, this is equivalent to $\mathscr{D}_k$ being a V-ring (5.21).

Let A be a differential ring (commutative) with derivation δ and $\mathscr{D}_A$ the ring of differential polynomials. Each element

$$d = \sum_{i=0}^{n} a_i \delta^i \in \mathscr{D}_A$$

determines a linear operator on A defined by

$$\alpha \mapsto \sum_{i=0}^{n} a_i \delta^i(\alpha) = d(\alpha).$$

To say that $\beta \in \operatorname{im} d$ is equivalent to saying that the n-th order linear differential equation (l.d.e.) $d(x) = \sum_{i=0}^{n} a_i \delta^i(x) = \beta$ has a solution in A, with $\ker d$ = solution space of $d(x) = 0$.

More generally, let $d_{ij} \in \mathscr{D}_A$, $i = 1, \ldots, m, j = 1, \ldots, n$ and $D = (d_{ij}) \in \mathrm{Mat}_{m \times n}(\mathscr{D}_A)$. D defines a map (also denoted D)

$$A^n \to A^m$$

as follows:

$$D\left[\begin{pmatrix} a_1 \\ \vdots \\ a_n \end{pmatrix}\right] = \begin{pmatrix} \sum_{j=1}^{n} d_{1j}(a_j) \\ \vdots \\ \sum_{j=1}^{n} d_{mj}(a_j) \end{pmatrix}.$$

(Note that we are viewing elements of the $(\mathscr{D}_A, A)$-bimodule A^n as column vectors of length n.)

For $\begin{pmatrix} b_1 \\ \vdots \\ b_m \end{pmatrix} \in A^m$, there is an associated system of linear differential equations

$$\sum_{j=1}^{n} d_{ij}(x) = b_i \quad i = 1, \ldots, m \tag{1}$$

or more succinctly

$$D(x) = b.$$

Naturally we would like to know when system (1) has a solution. By a solution to (1) we mean any column vector $\begin{pmatrix} a_1 \\ \vdots \\ a_n \end{pmatrix}$ such that

$$D\left[\begin{pmatrix} a_1 \\ \vdots \\ a_n \end{pmatrix}\right] = \begin{pmatrix} b_1 \\ \vdots \\ b_m \end{pmatrix}.$$

To that end observe that D induces D^*

$$D^* : {\mathscr{D}_A}^m \to {\mathscr{D}_A}^n \text{ defined by } (d_1, \ldots, d_m) \mapsto (d_1, \ldots, d_m) \cdot D$$

(ordinary matrix multiplication on the right by D). Here, we have agreed, for functorial reasons, to consider elements of the module ${\mathscr{D}_A}^m$ as row vectors having length m.

Let $C = \operatorname{cok} D^*$. Then

$${\mathscr{D}_A}^m \xrightarrow{D^*} {\mathscr{D}_A}^n \longrightarrow C \longrightarrow 0$$

is exact. Applying the functor $\hom_{\mathscr{D}_A}(-, A)$ we get (after some obvious identifications) the exact sequence

$$0 \longrightarrow \hom_{\mathscr{D}_A}(C, A) \longrightarrow A^n \xrightarrow{D} A^m.$$

Suppose $g \in A^m = \hom_{\mathscr{D}_A}(\mathscr{D}_A{}^m, A)$ and $D(f) = g$ for some $f \in A^n$, that is, f is a solution of the system $D(x) = g$. Then, clearly, g induces the zero element of $\hom_{\mathscr{D}_A}(K, A)$ where $K = \ker D^*$. For

$$\begin{array}{ccccccc} 0 & \longrightarrow & K & \longrightarrow & \mathscr{D}_A{}^m & \xrightarrow{D^*} & \mathscr{D}_A{}^n \\ & & & & \downarrow{\scriptstyle g} & & \downarrow{\scriptstyle f} \\ & & & & A & \xrightarrow{1} & A \end{array}$$

commutes and has exact rows.

Finally, let $J = \{g \in A^m \mid g$ induces the zero element of $\hom_{\mathscr{D}_A}(K, A)\}$. Then, clearly,

$$J/I \approx \operatorname{Ext}^1_{\mathscr{D}_A}(C, A) \quad \text{where} \quad I = \operatorname{im} D.$$

Thus, we can think of the extensions of A by C, $\operatorname{Ext}^1_{\mathscr{D}_A}(C, A)$ as obstruction classes to the solvability of $D(x) = g$, their triviality equivalent to solvability. Similarly, $\hom_{\mathscr{D}_A}(C, A)$ measures the number of different solutions. If $\operatorname{Ext}^1_{\mathscr{D}_A}(C, A) = 0$ for all finitely generated left $\mathscr{D}_A$-modules C, then each 'consistent system' $D(x) = g$ $(g \in J)$ has a solution in A and conversely. This condition is of course equivalent to the injectivity of A.

When $A = k$ is a field, k as a left $\mathscr{D}_k$-module is clearly simple and hence, there is an intimate connection between the 'real world' and the injectivity of certain simple modules. The next proposition is a trivial consequence of 3.7 and will be used in the proof of the main results of this section.

5.20 PROPOSITION *Let $M, N \in \mathscr{D}$-mod, M_R flat and ${}_{\mathscr{D}}N$ injective. Then $\hom_R(M, N)$ is an injective differential module.* □

5.21 THEOREM *The following are equivalent for $\mathscr{D} = \mathscr{D}_k$:*

(1) *${}_{\mathscr{D}}k$ is injective;*

(2) *each $M \in \mathscr{D}$-mod which is finite dimensional over k is injective in $\mathscr{D}$-mod;*

(3) *each linear differential equation has a solution in k.*

Moreover, ${}_{\mathscr{D}}k$ is the unique simple left $\mathscr{D}$-module up to isomorphism iff for each homogeneous l.d.e. has a nontrivial solution in k.

Proof (1) $\Leftrightarrow$ (3) and (2) $\Rightarrow$ (1) are clear.

(1) $\Rightarrow$ (2): Let M be finite dimensional and $M^* = \hom_k(M, k)$. Clearly, $M^*{}_k$ is flat and since ${}_{\mathscr{D}}k$ is injective, M^{**} is injective in $\mathscr{D}$-mod by 5.20.

However, since $\dim_k M < \infty$, $M \approx M^{**}$ in $\mathscr{D}$-mod by 3.8, completing the proof of the equivalence of (1) and (2).

Since $\mathscr{D}$ is a pli-domain, to show that ${}_{\mathscr{D}}k$ is unique, it suffices to show that

$$\hom_{\mathscr{D}}(\mathscr{D}/\mathscr{D}d, k) \neq 0$$

for all $d \in \mathscr{D}, d = \sum_{i=0}^{n} a_i\delta^i$. Consider the homogeneous linear differential equation

$$\sum_{i=0}^{n} a_i\delta^i(x) = 0$$

and let α be a nontrivial solution. Clearly, the map $f: \mathscr{D}/\mathscr{D}d \to k$ defined by $r + \mathscr{D}d \mapsto r \cdot \alpha$ is $\mathscr{D}$-linear and nonzero.

Conversely, if ${}_{\mathscr{D}}k$ is unique and $\sum_{i=0}^{n} a_i\delta^i(x) = 0$ is an arbitrary homogeneous linear differential equation, set $d = \sum_{i=0}^{n} a_i\delta^i$ and let $f: \mathscr{D}/\mathscr{D}d \to k$ be nonzero (f exists since ${}_{\mathscr{D}}k$ is unique). If $(1 + \mathscr{D}d)f = \alpha \neq 0$, $d(\alpha) = 0$ and hence, we have found the desired nontrivial solution of $d(x) = 0$. □

Constructing fields k satisfying condition (3) is quite easy (at least in theory). For example, if k is universal (cf. Cozzens [70]), then condition (3) is trivially satisfied. Thus, we have our first nontrivial examples of simple Noetherian V-domains which are not fields. Note that each cyclic $\mathscr{D}$-module, not isomorphic to $\mathscr{D}$, is injective. This is because such a module is necessarily finite dimensional over k (this also follows from 5.19(1)).

Analogous results hold for partial differential rings (see Cozzens–Johnson [72]). Of course, these rings can have any prescribed global dimension. Hence, nontrivial examples of V-domains exist in all dimensions.

A class of simple Noetherian V-domains with similar properties can be constructed via the twisted, finite Laurent polynomials $L = L_\rho(k)$ defined on p. 51. By considering difference equations of the form $\sum_{i=0}^{n} a_i\rho^i(x) = b$, ρ the automorphism of k, we obtain results analogous to Theorem 5.21. We shall refer to this Theorem by 5.21*. However, in this case, it is quite easy to give rather explicit examples of fields satisfying (3) of 5.21*.

5.22 EXAMPLE Let k be a field of characteristic p and $\rho : k \to k$ be defined by $\rho(\alpha) = \alpha^p$ for all $\alpha \in k$ (the Frobenius map). Clearly, ρ is an automorphism of k, having infinite period whenever k, is, say, separably closed. In this context, linear difference equations are polynomials of the form

$$\sum_{i=0}^{n} a_i X^{p^i} \quad a_i \in k. \tag{2}$$

By 5.21*, if each polynomial of the form

$$\sum_{i=0}^{n} a_i X^{p^i} = b \quad a_i, b \in k,$$

has a root in k, then L is a V-ring. Moreover, in order to have a unique simple L-module, each polynomial of type (2) containing at least two terms must have a nontrivial root in k. By letting $k = \bar{Z}_p$, these properties are obviously satisfied by algebraic closure of $\bar{Z}_p$.

Recently, Osofsky [71] showed that by suitably modifying the field k in Example 1, examples of simple Noetherian V-domains could be obtained with infinitely many nonisomorphic simple modules in contrast to the above example which has only one. In order to make all of this seem plausible, we shall exhibit another interesting property of the ring L which at the same time enables us to precisely determine the number of 1-dimensional simple L-modules.

Let $\mathscr{S}(1)$ denote the set of all isomorphism classes of one-dimensional (k-dimension, that is) L-modules. If M is one-dimensional, (M) will denote the class of M in $\mathscr{S}(1)$. For M, M' and M'' all one dimensional, $M \otimes_k M'$ is an L-module and clearly, $\dim_k(M \otimes_k M') = 1$. Moreover, it is quite easy to show that $M \otimes_k M' \approx M' \otimes_k M$ and $(M \otimes_k M') \otimes_k M'' \approx M \otimes_k (M' \otimes_k M'')$ as L-modules (of course the L-action is that defined in Chapter 3). Since $(k) \in \mathscr{S}(1)$ and $k \otimes_k M \approx M$ in L-mod, the operation $(M) \circ (M') = (M \otimes_k M')$ on $\mathscr{S}(1)$ is commutative, associative, and has identity (k). Thus, $(\mathscr{S}(1), \circ)$ is a commutative monoid. More can be said:

5.23 PROPOSITION *For $M \in L$-mod, $M^* = \hom_k(M, k)$*

$$M^* \otimes_k M \overset{\phi}{\approx} \hom_k(M, M) \textit{ as } L\textit{-modules,}$$

where ϕ is defined by

$$\phi(f \otimes m)[m'] = f(m')m$$

for all $f \in M^*$, $m, m' \in M$. □

It follows immediately from Proposition 5.23 that $(\mathscr{S}(1), \circ)$ is in fact, a group if $(M)^{-1}$ is defined to be (M^*).

5.24 THEOREM *Let k^* be the multiplicative group of units of the field k and define $\phi : k^* \to k^*$* by $\phi(\alpha) = \alpha^{-1}\rho(\alpha)$. *Then*

$$\mathscr{S}(1) \approx k^*/\text{im } \phi.$$

Proof Let $(M) \in \mathscr{S}(1)$ and $m \in M$. Then $\rho(m) = \alpha(m)m$ for a unique $\alpha(m) \in k$. Since $M = km$ for some $m \in M$,

$$\alpha(\beta m)\beta m = \rho(\beta m) = \rho(\beta)\rho(m) = \rho(\beta)\alpha(m)m.$$

Thus, $\alpha(\beta m) = \alpha(m) \bmod \text{im } \phi$. Hence, the correspondence $(M) \mapsto \alpha(m) \text{ im } \phi$ is a map $\psi : \mathscr{S}(1) \to k^*/\text{im } \phi$. The verifications that ψ is a surjective group homomorphism are routine and will be left to the reader.

Injectivity of ψ: Suppose $(M) \in \mathscr{S}(1)$, $M = km$ and $\alpha(m) \text{ im } \phi = \psi((M)) = \xi^{-1}\rho(\xi)$ for some $\xi \in k$. By the above, for each $m' \in M$, $\alpha(m') = \beta^{-1}\rho(\beta)$ for some $\beta \in k$. Since $\rho(m)$ can be written in the form $\beta\rho(\beta)^{-1}m$, we see that the map $M \to k$

defined by $\beta m \mapsto 1$ is an L-linear isomorphism, implying that $(M) = (k)$. $\square$

To exhibit a field k where (1) of 5.21* is satisfied but where ${}_L k$ is not unique, we proceed as follows.

Let q be a prime integer and let k be any field with algebraic closure $\bar{k}$. Let

$$\mathscr{F} = \{L \mid L \text{ is a field, } k \subseteq L \subseteq \bar{k} \text{ and } l \in L \Rightarrow \text{degree } l \text{ over } k \text{ is relatively prime to } q\}$$

5.25 DEFINITION A **q-field** over k is a maximal element of $\mathscr{F}$.

The utility of q-fields stems from

5.26 PROPOSITION (Osofsky [71]) *Let L be a q-field over k, and let $f \in L[X]$ have degree relatively prime to q. Then f has a root in L. If $g \in k[X]$ is irreducible and has degree divisible by q, then g has no root in L.* $\square$

To obtain the above mentioned example, let $k \supseteq F = Z_p(X)$ be a 2-field, $2 \neq p$, $\rho : k \to k$ defined by $\alpha \mapsto \alpha^p$ and $\{\pi_i \mid i \in \omega\}$ an infinite set of primes of $Z_p[X]$. Consider the cosets $\hat{\pi}_i = \pi_i \operatorname{im} \phi$ and $\hat{\pi}_j$ in $k^*/\operatorname{im} \phi$. We claim that they are distinct. For if $\hat{\pi}_i = \hat{\pi}_j$, then $\pi_i \pi_j^{-1} = \alpha^{p-1}$ for some $\alpha \in k^*$. Thus, the polynomial $f(Y) = \pi_j Y^{p-1} - \pi_i$ has a root in k^*. However, by Eisenstein's criterion, f is irreducible over $F[Y]$ and hence over $k[Y]$ by Proposition 5.26, since k is a 2-field and $2 \mid \deg f$. Thus, $\hat{\pi}_i \neq \hat{\pi}_j$ which implies that the cardinality of $(\mathscr{S}(1)) \geq \chi_0$.

6. *PCI-rings*

A 'good' theory of ring theory is one which generalizes the Wedderburn–Artin theorems for a ring R with radical N, and a 'good' theory of modules is one which generalizes the basis theorem for abelian groups. In this chapter we are concerned with the latter, in the setting of Dedekind domains, particularly the aspect which states that every finitely generated module M is a direct sum of finitely many ideals and cyclic modules. A 'conceptual' way to effect such a decomposition of M is to notice that M modulo its torsion submodule $t(M)$ is a torsion free module F embeddable in a free module P. Then, the canonical projections of $P \to R$ induce projections of F into R which furnish the requisite ideal summands. Moreover, $M/t(M)$ being projective implies that the torsion module $t(M)$ is a summand of M, and the cyclic decomposition for $t(M)$ is a consequence of the fact that R modulo any ideal $A \neq 0$ is an Artinian principal ideal ring, hence a uniserial ring over which every module is a direct sum of cyclics (as Köthe [35] showed).

Another way to obtain the cyclic decomposition is to observe that every proper cyclic module R/A is an injective R/A-module. Thus, every cyclic submodule 'of highest order' splits off!

We adopt this point of view in this chapter. However, instead of requiring that every proper cyclic right module R/A be injective modulo its annihilator ideal, we make the stronger hypothesis.

(right PCI) Every proper cyclic right R-module C is injective.

Proper cyclic means that C is cyclic but $C \not\approx R$.

We show that any right PCI-ring is either a semisimple ring, or a simple right semihereditary, right Ore domain. This reduces the entire theory to the simple right PCI-domains. As we have already observed, such rings occur freely in nature, constituting a broad class of the known simple domains.

In order to obtain these results however, we repeatedly use the following characterization of semisimplicity in terms of the class of cyclic modules, namely,

6.1 THEOREM (Osofsky) *A ring R is semisimple (Artinian) if and only if each cyclic right R-module is injective.* □

Thus, in terms of our development, the limiting case of a PCI-ring (hence V-ring) is a semisimple ring. For a short proof, see Osofsky [68].

Recall that if I is a right ideal of a ring R, then a right ideal K which is maximal in the set of all right ideals Q such that $I \cap Q = 0$ is called a **complement** right ideal, and K is said to be a **relative complement** of I. In this case, $I + K$ is an essential right ideal of R. Moreover, then, I is a relative complement of K if and only if I is a complement right ideal. In this case, I and K are said to be relative complement right ideals.

A ring A is **right neat** provided that the right singular ideal of A is zero. In this case, the injective hull $\hat{A}$ of A in mod-A is a right self-injective regular ring containing A as a subring. Moreover, for any right ideal I of A, a unique injective hull $\hat{I}$ of I in mod-A is contained in $\hat{A}$. Furthermore, $\hat{I}$ is a right ideal of $\hat{A}$, generated by an idempotent. Every annihilator right ideal of a right neat ring A is a complement, and every right ideal I of A is contained as an essential submodule in a unique complement $\bar{I}$ which is in fact $\hat{I} \cap A$ (see R. E. Johnson [51] or Faith [67*a*]).

6.2 LEMMA *If R is a right neat ring, and if I is a right ideal of R, then I is a complement right ideal of R if and only if R/I embeds in $\hat{R}$.*

Proof The complement right ideals of $\hat{R}$ are those of the form $e\hat{R}$, for some idempotent $e \in \hat{R}$, and those of R are of the form $e\hat{R} \cap R$. Thus, if $U = e\hat{R} \cap R$, then the isomorphism

$$R/e\hat{R} \cap R = R/U \approx eR$$

embeds R/U into $eR \subseteq \hat{R}$. Conversely, if $f: R/U \to \hat{R}$, is an embedding, and if $f(1 + U) = y$, then $U = y^{\perp} \cap R$. Write $\hat{R}y = \hat{R}(1 - e)$, for some idempotent e. Then $y^{\perp} = e\hat{R}$, and hence, $U = e\hat{R} \cap R$. □

6.3 THEOREM *Let A be a right PCI-ring. Then, either A is a regular ring, or else A is a simple ring. Moreover, A is a right V-ring, and every indecomposable injective right A-module is either simple or isomorphic to $\hat{A}$.*

Proof By Brown–McCoy [50], A has a maximal regular ideal M, and A/M has no regular ideals $\neq 0$ when $A \neq M$. If I is any ideal $\neq 0$, then every cyclic module C of A/I is a proper cyclic module over A, hence C is an injective module over A, and therefore injective over A/I. Then, 6.1 implies that A/I is semisimple Artinian. Hence, if A is not regular, then $M = 0$. Since every simple right module is injective, A is a right V-ring, and so by 5.15 a nonzero ideal B contains no right ideal isomorphic to A. Then, for any $y \neq 0$ in B, yA is injective, hence a summand of B generated by an idempotent. This implies that B is a regular ideal, contrary to the assumption that $M = 0$. Thus, A must be simple when A is not regular. If E is an indecomposable injective right A-module, then $E = \hat{C}$ for any nonzero cyclic submodule C. Thus, if $C \approx A$, then $E \approx \hat{A}$, and if $C \not\approx A$, then C is injective, whence $E = C$ is simple. □

6.4 PROPOSITION *A right PCI-ring A is right neat. Moreover, A has a nonsimple indecomposable injective right module E if and only if A is a right Ore domain, not a field. In this case, A is a simple ring.*

Proof Simple rings and regular rings are right neat, so the first sentence follows from 6.3. Since, by the remarks preceding 6.3, $\hat{A}$ is a regular ring, then $\hat{A}$ is indecomposable as a right A-module if and only if $\hat{A}$ is a field. In this case, A is an integral domain, and $\hat{A}$ is the right quotient field of A. Since A_A is not simple, then $A \neq \hat{A}$.

This proves one part of the second sentence, and the converse part is trivial. Then A is either regular or simple by 6.3 and the former can hold only if A is a field. Thus, A is simple in either case. □

6.5 PROPOSITION *A right-PCI domain R is right semihereditary, that is, each finitely generated right ideal of R is projective.*

Proof Let P be any nonzero projective right ideal, and let $a \in R$. If $C = aR + P$ is such that $C/P \approx R$, then there exists $x \in R$ such that $x^{\perp} = \{r \in R \mid ar \in P\}$. Since R is a domain, this would imply that $aR \cap P = 0$, and then $C = aR \oplus P$ is projective. If $C/P \not\approx R$, then C/P is injective, hence a summand of R/P. Let D be a right ideal of R containing P such that D/P is a summand of R/P complementary to C/P. Then there is an exact sequence

$$0 \to P \to C \oplus D \to R \to 0,$$

since $R = C + D$, and $P = C \cap D$. Thus, $C \oplus D \approx R \oplus P$ is projective, and hence C (also D) is projective. An obvious induction on the number of elements required to generate a right ideal shows that any finitely generated right ideal of R is projective, so R is right semihereditary. □

The next several results are preparation for the main results stated in Propositions 6.11 and 6.12.

6.6 LEMMA *A right ideal A of R is either semisimple, or else A contains an essential right ideal A' of R with $A' \neq A$. Moreover, A' is not a complement right ideal of R.*

Proof A module is semisimple iff it has no proper essential submodules. Thus, if A is not semisimple, there is an essential submodule $A' \neq A$. Let B be a relative complement of A' in R. If A' were a complement right ideal, then A' would be a relative complement of B, in which case $A \cap B \neq 0$, and $(A \cap B) \cap A' = 0$ contradicting the essentiality of A' in A. □

6.7 LEMMA *Let I and K be two right ideals of a ring R such that $I \cap K = 0$, and $I + K$ is essential. Then, the injective hull of $R/I \oplus R/K$ contains $\hat{R}$ as a summand.*

Proof Clearly, $R/I \oplus R/K$ contains the direct sum $I \oplus K$, which is essential in R, hence the injective hull contains the summand $\hat{R}$. □

6.8 PROPOSITION *Let R be a right PCI-ring. Then, either R is a right Ore domain, or else there is an exact sequence $R^2 \to \hat{R} \to 0$, or else R has nonzero right socle.*

Proof By 6.4, R is a right neat ring, with injective hull $\hat{R}$ which is a regular ring containing R as a subring. If every nonzero right ideal of R is essential, then $\hat{R}$ is a field, in which case R is a right Ore domain. Otherwise, there are right ideals I and K, neither of which are 0 or R, such that $I + K$ is essential and $I \cap K = 0$. If I' and K' are essential submodules of I and K, respectively, then $I' + K'$ is an essential right ideal of R, and by 6.7, the injective hull of $R/I' \oplus R/K'$ contains $\hat{R}$ as a summand. Now $R/I' \approx R$ would imply that I' is a complement right ideal of R by 6.2. However, by 6.6 I' is a complement right ideal essential in I only if $I' = I$. Thus, by 6.6 either I or K is semisimple, or else $R/I' \oplus R/K'$ is an injective module containing $\hat{R}$ as a summand. In the latter case, there is an exact sequence $R^2 \to \hat{R} \to 0$, as asserted. □

A right R-module E is **completely injective** provided that every factor module E/K is injective.

6.9 LEMMA *If R is a right PCI regular ring, and if e is an idempotent such that eR is completely injective (e.g., if e is contained in a nonzero ideal $S \neq R$), then eRe is a semisimple ring, and eR contains a minimal right ideal of R.*

Proof First, the canonical map $eR \otimes_R Re \to eRe$ is an isomorphism. To see this, let $\{a_i\}_{i=1}^n$ and $\{b_i\}_{i=1}^n$ be finitely many elements of R such that $\sum_{i=1}^n ea_ib_ie = 0$. Then:

$$\sum_{i=1}^n ea_i \otimes b_ie = \left(\sum_{i=1}^n ea_ib_ie\right) \otimes e = 0.$$

Henceforth, let $Q = eRe$, and let $M = Q/K$ be a cyclic right Q-module. Since Q is a regular ring along with R, then every

Q-module is flat, and therefore the canonical exact sequence $0 \to K \to Q \to M \to 0$ implies exactitude of

$$0 \to K \underset{Q}{\otimes} eR \to Q \underset{Q}{\otimes} eR \to M \underset{Q}{\otimes} eR \to 0.$$

Since $Q \underset{Q}{\otimes} eR \approx eR$ as a (Q, R)-bimodule, this implies that $C = M \underset{Q}{\otimes} eR$ is an epimorphic image of eR in mod-R, and hence is injective. (If e is contained in a nonzero ideal $S \neq R$, then $eR/L \approx R$, for some right ideal L, would imply via projectivity of R that eR contains a submodule $\approx R$, contrary to 5.15 which states that S contains no right ideals $\approx R$. Thus, eR is completely injective in this case.)

In order to show that M is injective in mod-Q, it suffices to show that M is isomorphic to a summand of any over-module. Since $C = M \underset{Q}{\otimes} eR$ is injective in mod-R, if Y is any right Q-module containing M, then the induced inclusion $C \to Y \underset{Q}{\otimes} eR$ splits. Write $Y \underset{Q}{\otimes} eR = C \oplus X$, for some right R-module X. Now,

$$(Y \underset{Q}{\otimes} eR) \underset{R}{\otimes} Re \approx (C \underset{R}{\otimes} Re) \oplus (X \underset{R}{\otimes} Re).$$

Since $eR \underset{R}{\otimes} Re \approx Q$ canonically, it follows that

$$C \underset{R}{\otimes} Re \approx (M \underset{Q}{\otimes} eR) \underset{R}{\otimes} Re \approx M \underset{Q}{\otimes} (eR \underset{R}{\otimes} Re)$$
$$\approx M \underset{Q}{\otimes} Q \approx M.$$

Thus, M is isomorphic in mod-Q to a summand of $Y \approx (Y \underset{Q}{\otimes} eR) \underset{R}{\otimes} Re$ and so M, hence every cyclic right Q-module, is injective. Then, Q is semisimple. Let I be a minimal right ideal of Q. Since Q is semisimple, $I = fQ$ for an idempotent $f \in Q$, and then $fQf = fRf$ is a field $\approx \operatorname{End} fQ_Q$. Since R is semiprime, this implies that fR is a minimal right ideal of R contained in eR (Jacobson [64*a*], p. 65). □

6.10 Lemma *If R is a right PCI-ring, and if S is any ideal $\neq 0$, R, then S contains any right ideal I for which there is an* *isomorphism* $f: R/I \to R$. *Moreover,* $I = x^{\perp}$ *for* $x = f(1 + I)$, *and* $xy = 1$, *for some* $y \in R$.

Proof Let $f: R/I \to R$ be an isomorphism. Then, $I = x^{\perp}$, where $x = f(1 + I)$, and $xR = R$. Thus, $xy = 1$ for some $y \in R$. Since every cyclic right R/S-module is injective, then R/S is semisimple Artinian by 6.1. Therefore, every one-sided inverse in R/S is two-sided, and consequently, $yx = 1 - s$, for some $s \in S$. Thus, if $xa = 0$, then $a = sa \in S$. This proves that $I = x^{\perp}$ is contained in S. □

6.11 Proposition *If* R *is a right PCI-ring, and if* R *is regular, then* R *is semisimple.*

Proof Let S denote the intersection of all nonzero ideals of R. The proof falls into three cases:

(1) $S = 0$ but R is not simple.

By Lemma 6.10 if R/I is a cyclic right R-module isomorphic to R, then I is contained in every nonzero ideal of R. Thus, (1) implies that R/I is injective for every right ideal $I \neq 0$. If R has idempotents $e \neq 0$, $e \neq 1$, then $R/eR \approx (1 - e)R$ is injective, and, similarly, eR is injective, in which case, $R = eR \oplus (1 - e)R$ is injective. Thus, every cyclic right R-module is injective, so R is semisimple.

(2) $S \neq 0$.

First, R/S is semisimple, since every cyclic R/S-module is injective. Second, by Lemma 6.9 R has nonzero socle. The fact that $S = S^2$ is the least ideal of R implies that R is prime, and hence that S is the right socle of R. If V is any minimal right ideal of R, then V is injective, and hence $V\hat{R} = V$ is a right ideal of $\hat{R}$, and in fact, V is a simple right ideal of $\hat{R}$. Thus, right socle $\hat{R} \supseteq S$. Furthermore, if W is a minimal right ideal of $\hat{R}$, then $W \cap R = U$ is a minimal complement right ideal of R such that $W = U\hat{R}$. But every nonzero right ideal of R contains a minimal right ideal, and the latter is a minimal complement. Thus, U is a minimal right ideal of R, and, as indicated above, $U = U\hat{R}$. Hence, $W = U$ is a minimal right ideal of R. This proves that $S =$ socle $\hat{R}$. Therefore, S is an ideal of $\hat{R}$. If the socle S is finite, then $S = R$ is semisimple, and the proof is complete. Otherwise, $S \approx V^{(a)}$ for some infinite cardinal a, and hence, there is a right ideal K contained in S such that $K \approx V^{(a)} \approx S \approx K^2$. Then, $R/K \approx R$ would imply that $K = eR$ for an idempotent $e \in R$, in which case K, whence S, is a finite sum of injective simple modules. But this case has just been discarded. Thus, R/K is an injective module containing a submodule $K_1 \approx S$, and hence containing the injective hull $\hat{R}$ of S. Since $\hat{R}$ is then a summand of a cyclic module, $\hat{R}$ is cyclic, and hence there exists $x \in \hat{R}$ such that $\hat{R} = xR$. Let $a \in R$ be such that $xa = 1$. Since S is an ideal of $\hat{R}$, if $c \in R$, and if $ac \in S$, then $c = xac \in S$. This proves that $[a + S]$ is not a left zero divisor in R/S. Since R/S is semisimple, this means that $[a + S]$ is a unit of R/S, with inverse, say $[a' + S]$. Then, $aa' = 1 - s$, for some $s \in S$, and

$$xaa' = x(1 - s) = a'.$$

Thus, $x = xs + a'$. Since S is an ideal of $\hat{R}$, then $x \in R$, and therefore, $\hat{R} = xR \subseteq R$. Since every cyclic module is therefore injective, R is semisimple in this case too.

(3) R is simple.

This case follows from the next proposition.

6.12 PROPOSITION *A simple right PCI-ring R is either semisimple, or else a right semihereditary simple domain.*

Proof Case (1). R contains a nonzero injective right ideal A. Since R is simple, then A is a generator of mod-R, which implies that $R = \hat{R}$ and hence, R is semisimple by 6.1.

Case (2). If (1) does not hold, then, since every simple module is injective, R has zero right socle. Then, by 6.8, either R is a right Ore domain, or else $R^2 \to \hat{R} \to 0$ is exact. In the former case, R is right semihereditary by 6.5. Hence, assume:

Case (3). $R^2 \to \hat{R} \to 0$ is exact. Since R has no injective right ideals $\neq 0$, every principal right ideal $aR \neq 0$ is isomorphic to R. Let $f: aR \to R$ be an isomorphism. Then, $f(a) = x$ is an element of R such that $xR = R$, and $x^\perp = a^\perp$. Let $y \in R$ be such that $xy = 1$. Then, $e = yx$ is idempotent. If $e \neq 1$, then $eR \approx R$ and $(1 - e)R \approx R$, hence $R \approx R^2$. This implies that $R \to \hat{R} \to 0$ is exact and $R \to R \oplus \hat{R} \to 0$ is exact. Since $C = R \oplus \hat{R}$ is thereby cyclic, then C is either injective, whence R is injective, or else $C \approx R$, whence R contains an injective right ideal $A \approx \hat{R}$. But, in either case, R is semisimple. This proves that $e = yx = 1$, whence that $x^\perp = a^\perp = 0$ for every nonzero $a \in R$. Then R is an integral domain, which must be simple and right semihereditary by 6.5. □

6.13 THEOREM *A right PCI-ring R is either semisimple, or a right semihereditary simple domain.*

Proof By 6.3, R is either regular, whence semisimple by 6.11 or else R is simple, in which case the last proposition applies. □

6.14 PROPOSITION *Let R be a right PCI-domain which is not right Ore, then:*

(1) *$\hat{R}$ is a finitely presented cyclic module.*

(2) *If I is any right ideal, then $R/I \approx \hat{R}$ iff I is a complement right ideal $\neq 0$, R. Furthermore, any complement right ideal I is finitely generated* (*by* 3 *elements*).

(3) *Any finitely generated torsion right module M is injective, and σ-cyclic* (*that is, a finite direct sum of cyclic modules*).

(4) *If I is a right ideal, then there is a unique least complement $\bar{I}$ containing I, and $\bar{I}$ is the unique complement containing I as an essential submodule. Moreover* (*any finitely generated submodule of*) *$\bar{I}/I$ is cyclic and injective.*

(5) *If $x \in \hat{R}$, and $x \notin R$, then $xR = \hat{R}$ iff $xa = 1$ for some $a \in R$. Moreover, in this case, $J_a = x^\perp + aR$ is a finitely generated essential right ideal, and $R/J_a \approx \hat{R}/R$.*

Proof of (1) If I is any complement right ideal $\neq 0$, and if K is the relative complement of I, then K embeds canonically as an essential submodule of R/I. Since R/I is injective, then $R/I \approx \hat{K}$, so we must show that $\hat{K} \approx \hat{R}$. Now K contains a right ideal $aR \approx R$, and hence we have

$$\hat{R} \supseteq \hat{K} \supseteq \widehat{aR} \approx \hat{R},$$

so by a theorem of Bumby and Osofsky (see Bumby [65]) we have $\hat{K} \approx \hat{R}$ as desired. This will prove (1), once we complete the proof of (2) showing that I is finitely generated.

Proof of (2) Let aR be a nonzero cyclic right ideal which is not essential. Then R/aR is an injective module containing a submodule $X \approx \hat{R}$. Write $R/aR = Y/aR \oplus X$, for some module $Y \supseteq aR$. Then, $R/Y \approx X \approx \hat{R}$, so that Y is a complement right ideal by 6.2. Furthermore, Y/aR is cyclic, so that $Y = aR + bR$, for some $b \in R$, is finitely generated, hence projective. Therefore, $\hat{R}$ is finitely presented. If I is any complement right ideal $\neq 0, R$, then $R/I \approx \hat{R}$, by 1. Then Schanuel's lemma implies that $R \oplus Y \approx R \oplus I$, so I is generated by 3 elements. Conversely, if I is a right ideal such that $R/I \approx \hat{R}$, then I is a complement right ideal by Lemma 6.2. This proves (2).

Proof of (3) If $M = a_1R + \ldots + a_nR$ is finitely generated and torsion, then a_1R is injective, so $M = a_1R \oplus M_1$ for some submodule M_1. Now M_1 is generated by $a_2', \ldots, a_n'$, where b' is the image of any $b \in M$ under the canonical projection $M \to M_1$. By induction, M_1, hence M, is a direct sum of finitely many injective cyclic modules. Therefore, M is injective.

Proof of (4) By the remarks preceding 6.2, in any right neat ring R, any right ideal has just one injective hull $\hat{I}$ (=maximal essential extension) contained in $\hat{R}$. Then $\bar{I} = \hat{I} \cap R$ is the unique complement containing I as an essential submodule. Moreover, since any least complement containing I is essential over I, then $\bar{I}$ is the unique least complement containing I. Any finitely generated submodule of $\bar{I}/I$ is injective by (3), hence a summand of R/I, and therefore cyclic.

Proof of (5) If $xR = \hat{R}$, then $\exists a \in R$ with $xa = 1$. Conversely, if $xa = 1$ with $x \in \hat{R}$, $a \in R$, $x \notin R$, then $x^{\perp} \neq 0$ since $x(1 - ax) = 0$. Thus, $xR \approx R/x^{\perp}$ is a proper cyclic, therefore injective. But $xR \supseteq R$ and therefore $xR = \hat{R}$.

Let $J = \{r \in R \mid xr \in R\}$. Clearly, $xJ = R$, and there is a (canonical) isomorphism $xR/xJ \approx R/J$ (sending $[xr + xJ] \mapsto r + J, \forall r \in R$). Since $xR = \hat{R}$, we have the desired isomorphism $\hat{R}/R \approx R/J_a$ provided we show that J coincides with $J_a = x^{\perp} + aR$. Clearly $J \supseteq x^{\perp} + aR$. Moreover, if $y \in J$, $y = (y - axy) + axy$ and $y - axy \in x^{\perp}$, $axy \in aR$ which shows that $J = x^{\perp} + aR = J_a$. □

Following from 6.14(3), we have a theorem reminiscent of the situation for hereditary Noetherian prime rings and Dedekind rings (cf. Kaplansky [52], Levy [63]).

6.15 COROLLARY *If R is a right PCI, right Ore domain, then every finitely generated right ideal is projective and generated by two elements. Furthermore assuming that R is left Ore, then any finitely generated module M is a direct sum of a finite number of right ideals of R, and cyclic injective modules.*

Proof Over a two-sided Ore domain, any finitely generated torsion free module F is embeddable in a free module of finite rank (see 2.36). Thus, since R is semihereditary by 6.5, $F = M/t(M)$ is projective, so the canonical map $M \to F$ splits. Thus, $M \approx F \oplus t(M)$. Moreover, F is isomorphic to a direct sum of right ideals. Finally, the proof of 6.14(3) suffices to establish that $t(M)$ is σ-cyclic, and injective. In particular, if I is any finitely generated right ideal, and if a is any nonzero element of I, then I/aR, being finitely generated torsion, is injective, and, being a summand of R/aR, is therefore cyclic. Thus, $I = bR + aR$ for some $b \in R$. □

The proof of 6.14(2) has the corollary:

6.16 COROLLARY *If R is a right neat ring, and if I is a right ideal such that $\hat{R}$ embeds in R/I, then there exists an element $b \in R$ such that $Y = bR + I$ is a complement right ideal of R.* □

6.17 THEOREM *A right PCI-domain is right Ore.*

Proof If R is not right Ore, then we may assume that $\hat{R} \approx R/I$ is a finitely presented cyclic module by 6.14(1), that is, that I is finitely generated. Write

$$0 \to I \to R \to \hat{R} \to 0 \tag{1}$$

exact. Now, any finitely generated left R-submodule M of $\hat{R}$ can be embedded in R (via a right multiplication by a nonzero element of R), and so $\hat{R}$ is a direct limit of projective modules, whence $\hat{R}$ is a flat left R-module (all of which is well-known; e.g., Sandomierski [68]). Thus, we have exactness of

$$0 \to I \otimes_R \hat{R} \to R \otimes_R \hat{R} \to \hat{R} \otimes_R \hat{R} \to 0. \tag{2}$$

Since $\hat{R}$ is left flat, and I is finitely generated, then $I \otimes_R \hat{R} \approx I\hat{R}$ is a finitely generated right ideal of $\hat{R}$, which by the remark in the proof of 6.14(1) is isomorphic to $\hat{R}$, whence a summand of $\hat{R}$.

This proves that $\hat{R} \underset{R}{\otimes} \hat{R}$ is isomorphic to a summand of $\hat{R}$, whence has zero singular submodule on the right. Now the kernel of the canonical map $\hat{R} \underset{R}{\otimes} \hat{R} \to \hat{R}$ is contained in the singular submodule, hence must equal 0. This proves that the embedding of R into $\hat{R}$ is a ring epic (Silver [67]). This, together with left flatness of $\hat{R}$, suffices to show that a right $\hat{R}$-module M is injective as a right $\hat{R}$-module iff it is injective as a right R-module. In particular, if M is a proper cyclic right $\hat{R}$-module, we conclude that M is injective, since M is a proper cyclic right R-module. Then, by Osofsky's theorem, $\hat{R}$ is semisimple, which can happen iff R is a field. This proves that R is right Ore. □

It is an open question whether or not a right PCI-domain R is right Noetherian. If it is, then every injective right module M is Σ-injective in the sense that any direct sum of copies of M is injective. It is known that to test for Σ-injectivity of M it suffices to test injectivity of $M^{(\omega)}$ (Faith [66]). Moreover, over a PCI-domain, every indecomposable injective module E is either simple, or $E \approx \hat{R}$ is the right quotient field of R. The latter is always Σ-injective by a theorem of Faith [66]. So, in order to test Σ-injectivity of indecomposable injectives, it suffices to consider just simple modules. This we do in 6.18 – 6.19 below.

6.18 Proposition *Let R be any ring, and let V be a simple injective right R-module which is not Σ-injective. Then, some cyclic right R-module R/J has essential socle $\approx V^{(\omega)}$, a direct sum of countably infinitely many V's.*

Proof As stated, an injective module M is Σ-injective if and only if $M^{(\omega)}$ is injective. Assuming that V is not Σ-injective, then $V^{(\omega)}$ has injective hull $E \neq V^{(\omega)}$. Since $V^{(\omega)}$ is essential in E, this implies that E is not semisimple, and hence there is a cyclic submodule xR of E which is not semisimple. Since E has essential socle, then xR has essential socle H. Since every finite direct sum of injective modules is injective, essentiality of H in xR implies that H does not have finite length. Since H is contained in $V^{(\omega)}$, this implies that $H \approx V^{(\omega)}$. □

The proof has the corollary.

6.19 COROLLARY *If R is a right PCI-ring, and if R is not right Noetherian, then there is a cyclic injective right R-module E with infinite essential socle.*

Proof By 6.13, R must be a domain, not a field. By 5.2, not every semisimple module is injective. Hence, there is a semisimple module M with $\hat{M}$ not semisimple, hence containing a cyclic submodule xR which is not semisimple. Then, xR has infinite socle, since any finite direct sum of simple modules is injective. Moreover, since R is not a field, soc $R = 0$, so $xR \not\approx R$, that is, xR is injective. □

6.20 PROPOSITION *Let R be a right PCI-ring, let I be a finitely generated essential right ideal, and let $E = R/I$. Then, $B = \text{End } E_R$ is a right self-injective regular ring.*

Proof It is clear that E is completely injective. If $b \in B$, then $C = bE$ is injective, hence a summand of E. Write $C = X/I$ for a right ideal X, and let Y/I be the complementary summand of E. Then Y/I is cyclic, so Y is finitely generated. This implies that $C \approx R/Y$ is finitely presented. If ker $b = K/I$, then there is an exact sequence $0 \to K \to R \to C \to 0$, hence by Schanuel's lemma, K is finitely generated. But, by induction, every finitely generated submodule of E is injective. Thus, ker b is essential only if $b = 0$. By Utumi's theorem (Appendix 1), this proves the proposition. □

6.21 COROLLARY *Let R be a right PCI-domain, with a unique simple right R-module V. Assume that R is not right Noetherian, and let I be the right ideal given by* 6.19 *such that $E = R/I$ is injective with infinite essential socle. Then I is not finitely generated.*

Proof $E = R/I$ is not semisimple, hence there exists a maximal right ideal $M \supseteq I$ such that M/I contains the socle of R/I. The socle S of $E \approx$ direct sum of copies of V, so there exists a nonzero map $R/M \to E$. Thus, M/I is an essential submodule of R/I, and is the kernel of an endomorphism f of E. Then, f is contained in the radical of End E_R, so the preceding proposition shows that I cannot be finitely generated. □

Since $B = \text{End } E_R$ has nonzero radical J, and $B/J \approx \text{End } S_R$, B/J is a full right linear ring.

6.22 Definition A ring R is a **right fir** provided that every right ideal of R is a free module, and every free right module has invariant basis number (IBN). Thus, an isomorphism $R^{(I)} \approx R^{(J)}$, for sets I and J, implies card I = card J.

6.23 Corollary *A right PCI-domain R is a principal right ideal domain iff R is a right fir.*

Proof Any right fir which is also right Ore is necessarily a pri-ring. For if I is any right ideal generated by a_1 and a_2, $a_1 \notin a_2R$, then since $a_1R \cap a_2R \neq 0$, $a_1x = a_2y \neq 0$ for $x, y \in R$. Hence, we obtain a nontrivial linear relation in I, namely $a_1x + a_2(-y) = 0$ contradicting freeness of I. So, $I = a_2R$.

The proof of the corollary is now clear. □

6.24 Proposition *If R is a right Noetherian right PCI-domain with right quotient field $Q \neq R$, then R is right hereditary and*

(1) *Q/R is semisimple;*

(2) *R embeds canonically in* End Q/R *under a map: $r \mapsto r'$, where r' sends $[x + R] \mapsto [rx + R]$;*

(3) *for any right ideal I, the factor module R/I is a semisimple module of finite length;*

(4) *any finitely generated torsion module is semisimple of finite length.*

Proof A right PCI-domain R is right semihereditary by 6.5, and hence Noetherian implies right hereditary.

(1) The module Q/R is a torsion module, and by 6.14(3) every finitely generated torsion module is injective.

Since R is right Noetherian, this implies that for any finitely generated submodule M of Q/R, every submodule is a summand, and hence that M is semisimple. It readily follows that Q/R is semisimple.

(2) Obvious. (Note that since R is simple, and the map is nonzero, the map is an embedding.)

Proofs of (3) and (4) are similar to that of (1) (cf. 6.14(3).

□

6.25 THEOREM *A left Ore right Noetherian right PCI-domain is left Noetherian.*

Proof Let

$$I_1 \subseteq I_2 \subseteq \ldots \subseteq I_n \subset R \tag{3}$$

be an ascending chain of finitely generated (hence projective) left ideals of R, $c \in I_1$ and D, the 2-sided quotient field of R. The set

$$I_i^{-1} = \{d \in D \mid I_i d \subseteq R\}$$

is a right R-submodule of D, canonically isomorphic to $I_i^* = \hom_R(I_i, R)$ since ${}_R I_i$ is essential. By applying $(\ )^{-1}$ to (3) we obtain

$$(Rc)^{-1} \supseteq I_1^{-1} \supseteq I_2^{-1} \supseteq \ldots \supseteq R.$$

Since $(Rc)^{-1} = c^{-1}R$, multiplication by c yields

$$R \supseteq cI_1^{-1} \supseteq cI_2^{-1} \supseteq \ldots \supseteq cR.$$

By 6.24(3), R/cR is Artinian. Thus for some $n > 0$, $cI_n^{-1} = cI_{n+j}^{-1}$ for all $j \geq 1$ and hence $I_n^{-1} = I_{n+j}^{-1}$. However, $I_n^{-1-1} = I_n$ for all n by projectivity of I_n and hence, $I_n = I_{n+j}$ for all $j \geq 1$. Thus, chain (3) terminates in finitely many steps. □

6.26 THEOREM (Boyle [74]) *A right and left Noetherian ring R is a right PCI-ring iff R is a right hereditary right V-domain. Furthermore, in this case, R is a left PCI-domain.*

Proof Since a two-sided Noetherian, right hereditary ring is left hereditary, (Jans [64], p. 58), the first statement follows trivially from 5.19(1). By 6.24(4), each finitely generated torsion right R-module is semisimple and hence, each finitely generated torsion left R-module is semisimple by 2.32. Consequently, each left ideal is a (finite) intersection of maximal left ideals. Thus, R is a left V-domain and hence, a left PCI-domain by 5.19.1. □

7. *Open problems*

Let $A = \text{End}\,_B U$ be a simple right Goldie ring, with U_A a uniform right ideal finitely generated projective and faithful over the right Ore domain B, with $T = \text{trace}\,_B U$ a least ideal of B. (Some of these requirements are redundant.)

(1) *Is A similar to a domain R?*

This can happen iff A contains a (finitely generated) projective uniform right ideal W.

(2) *Is $A \approx R_n$ a full $n \times n$ matrix ring over a domain R?*

Clearly yes to (2) implies yes to (1).

(3) *Does A have nontrivial idempotents when A is not a domain?*

If $e = e^2 \in A$ is nontrivial, then eA is a (B, A)-progenerator, where $B = eAe \approx \text{End}\, eA_A$. Suppose that (3) has an affirmative answer. Since B has Goldie dimension less than that of A, an induction on the Goldie dimension shows that there exists $e \in A$ such that $B = eAe$ is a domain. Then A is similar to a domain, since eA is a (B, A)-progenerator. Thus, yes to (3) implies yes to (1).

(*Note added in proof:* A. Zalesski has announced an example giving a negative answer.)

(4) *Let R be a right Ore domain with a finitely generated projective left module W such that $T = \text{trace}\,_R W$ is a least ideal of R. Can W be chosen such that W is indecomposable of rank > 1?*

(The rank of W is defined to be to the dimension over D of the left vector space $D \bigotimes_R W$, where D is the right quotient field of R.)

(5) *If R is a simple right Ore domain, is every finitely generated projective left R-module W a direct sum of modules of rank 1?*

By induction on the ranks, one can see that the answer to (5) is yes iff that to (4) is no for simple R. (In this case trace ${}_R W = R$ for every W.) Moreover, yes for (5) holds iff yes to (3) holds.

(6) *Assume that the answer to* (2) *is yes, and let B be a simple right Ore domain. If W is any finitely generated projective module over B, then there exists a projective module U of rank* 1 *over B, and an isomorphism $W \approx U^n$, for some integer $n \geq 1$.*

For then $R_n = A = \text{End}\,{}_B W$ is a simple right Goldie ring, and $U = We_{11}$ has the desired property, where e_{11} is the matrix unit of R_n (in the usual notation). Clearly U has rank 1 because $\text{End}\,{}_B U \approx e_{11}Ae_{11} \approx R$ is right Ore domain. Moreover

$$W = \sum_{i=1} We_{ii} \approx We_{11} \oplus \ldots \oplus We_{11} \approx U^n.$$

(7) *Does there exist a right V-domain with a prescribed* (*finite*) *number of simple right modules?*

In Chapter 5, we presented Cozzens' examples of simple V-domains, each having a unique simple module (on either side). We also constructed Osofsky's example of a V-domain with countably many simple modules. But so far we have neither an example of a right V-domain which is not a left Ore domain, nor one having a finite number n of simple modules for any $n > 1$.

(8) *Determine new examples of simple Noetherian* (*domains*) *rings unlike those described in Chapter* 3.

(9) *Is the ultraproduct* (*defined presently*) *of countably many copies of a simple Noetherian V-domain such as that described in Chapter* 5 *necessarily a V-domain? a PCI-domain?*

Recall that a **filter** $\mathscr{F}$ of subsets of a set I is a family of subsets of I satisfying

(*a*) $\varnothing$, the empty set, is not in $\mathscr{F}$,

(*b*) $S_1, S_2 \in \mathscr{F} \Rightarrow S_1 \cap S_2 \in \mathscr{F}$,

(*c*) $S \in \mathscr{F}, S \subseteq T \subseteq I \Rightarrow T \in \mathscr{F}$.

An **ultrafilter** is defined to be a maximal filter which obviously exists by Zorn's lemma.

Suppose $\{R_\alpha : \alpha \in I\}$ is a family of rings and $\bar{R} = \prod_{\alpha \in I} R_\alpha = \{f \mid f: I \to \bigcup_{\alpha \in I} R_\alpha \text{ and } f(\alpha) \in R_\alpha \forall \alpha\}$. If $\mathscr{F}$ is any filter on I, we define $f \equiv g (\text{mod } \mathscr{F})$ iff $\{\alpha \mid f(\alpha) = g(\alpha)\} \in \mathscr{F}$. Clearly, $\bar{I} = \{f \in \bar{R} \mid f \equiv 0 (\text{mod } \mathscr{F})\}$ is a two-sided ideal of $\bar{R}$. The factor ring $\bar{R}/\bar{I}$ is usually denoted $\prod_{\alpha \in I} R_\alpha/\mathscr{F}$ and is called an **ultraproduct** whenever $\mathscr{F}$ is an ultrafilter.

It is known that ultraproducts preserve certain properties of the rings R_α, namely, those which can be described by first order statements (see, e.g., Robinson [63]). For example, the properties of being a domain, a field, an Ore domain, semihereditary are all first order properties; being Noetherian and hereditary however, are not first order properties.

If R_α is the domain described above for all $\alpha \in \omega$, set $U = \prod R_\alpha/\mathscr{F}$. U is necessarily a non-Noetherian, semihereditary Ore domain. Moreover, U is simple since the following first order property, equivalent to simplicity, holds in each R_α: $\forall a \in R_\alpha$, $\exists x, y \in R_\alpha$ such that

$$xa + ay = 1.$$

(10) *Find an example of a right Noetherian right V-(PCI-) domain which is not a left V-(PCI-) domain.*

Here is a possible method for generating the desired nonsymmetric examples. Let δ be a ρ-derivation of a noncommutative field k with ρ-nonsurjective. Extend $\delta(\rho)$ to a field $\bar{k} \supset k$ in such a manner that

(*a*) every linear differential equation in δ has a solution in $\bar{k}$,

(*b*) ρ is still nonsurjective.

Clearly, if this is possible, the ring of ρ-differential operators with coefficients in $\bar{k}$ will be the desired one-sided example (see Theorem 5.21).

(11) *Find an example of a non-Noetherian PCI-domain.*

(12) *Find a left Noetherian right Ore domain which is not right Noetherian* (see example 2, Chapter 4).

(13) *Is the tensor product of two simple V-rings $A \otimes_C B$ over the center C of A necessarily a V-ring, e.g., is it true for fields? e.g., if D is a field with center C, is $D \otimes_C C(X)$ a V-ring? (If D/C is an algebraic algebra, then $D \otimes_C C(X)$ is a field whenever $D[X] = D \otimes_C \hat{C}[X]$ is not a primitive ring.)*

A ring R is a **right QI-ring** if every quasi-injective right R-module is injective.

The examples of Noetherian V-rings of Cozzens [70] are all QI-rings, and principal (right and left) ideal domains. (See Chapter 5.) These domains are thus hereditary. Boyle [74] proved that Noetherian hereditary V-rings are QI, and conjectured that every right QI-ring is right hereditary. Now a theorem of Webber [70] as generalized by Chatters [71] states that any (2-sided) Noetherian hereditary ring satisfies the **restricted right minimum condition, RRM**: R satisfies the d.c.c. on right ideals containing any essential right ideal.

Thus, for (right and left) QI-rings, RRM is a necessary condition for the truth of Boyle's conjecture.

A ring R has **right Krull dimension** K-dim ≤ 1 if each properly descending chain

$$A_1 \supset A_2 \supset \ldots \supset A_n \supset$$

of right ideals of R, such that for each $i = 1, 2, \ldots$ the right R-module A_i/A_{i+1} is not Artinian, has only finitely many terms.

A structure theorem of Michler–Villamayor [73] classifies right V-rings of right Krull dimension (K-dim) ≤ 1 as finite products of matrix rings over right Noetherian, right hereditary simple domains, each of which are restricted semisimple. These rings are thus right hereditary, verifying Boyle's conjecture for these rings and in particular for RRM-rings.

This introduces the problem:

(14) *Prove Ann Boyle's conjecture.*

For an extension of the Michler–Villamayor result, using different methods, consult Faith [74*b*], who proves that any right QI simple but nonsemisimple ring R possesses an injective nonsimple indecomposable right module E such that every factor module of any direct sum of copies of E is injective, and any proper factor module of E is semisimple. Moreover, if every nonsimple indecomposable injective right module E has this property, then R is right hereditary (Faith [74*b*]). A right QI-ring has the property that the endomorphism ring of an indecomposable injective module is a field. This property for a right Noetherian ring has been characterized by Faith [74*a*].

Afterword

The most important class of simple rings is the category of fields, denoted FIELDS. These are the rings in which every nonzero element is a unit, and, like the quaternions, may be noncommutative. They have also been called division rings, but we retain this terminology only if R is a finite dimensional algebra over its center, in which case we say division algebra. The literature on division algebras is vast, and we refer the reader either to Albert [39], or to Deuring [48] for an account up to about 1936. Recently, Amitsur [72] solved the question of the existence of division algebras which are not crossed products in the negative, so the reader may refer to this paper for new perspectives.

Steinitz extensions

Steinitz [48] determined that any commutative field F can be obtained as an extension of the prime subfield P by two intermediate extensions: first, a purely transcendental extension T/P (meaning that T is generated by a set $\{x_i\}_{i \in I}$ of elements any finite subset of which generates a rational function field over P in those variables); second, F is an algebraic extension of T in the sense that every element $y \in F$ satisfies a nonzero polynomial over T.

Cartan–Brauer–Hua

Unfortunately, a Steinitzian simplicity cannot hold for a noncommutative field F, since a very elementary theorem, the Cartan–Brauer–Hua theorem (see Jacobson [64*a*]), implies that F is generated as a field by the conjugates of any noncentral subset. Therefore, if some element y of F is transcendental, that is, not algebraic, over its center, then F is generated by

$\{x^{-1}yx\}_{x \in F^*}$, where F^* denotes the units group of F, so F is then generated by transcendental elements. Similarly, if F is not purely transcendental, that is, if some noncentral element y of F is algebraic over its center, then F is generated by the conjugates of y, hence by algebraic elements!

If, as we have indicated, the category FIELDS admits no quick classification in either of the two classical modes, i.e., the classification of all rational division algebras as 'cyclic division algebras' as introduced by L. E. Dickson, or in Steinitzian style, then how can we expect a rapid purview of the much larger class of simple rings?

Wedderburn–Artin

Of course, great simplifications are possible if we choose the Wedderburn–Artin gambit: any simple ring R with descending chain condition on right ideals, that is, any simple right Artinian ring, is a full $n \times n$ matrix ring D_n over a uniquely determined field D. This 'reduces' the structure of R to that of an 'underlying' field (i.e., *forget* the field!).

The power of the Wedderburn–Artin theorem lies in the fact that the structure of any module over a semisimple ring is transparent, namely, a direct sum of simple modules. This fact makes it *the* fundamental tool in the structure of right Artinian rings, since if J denotes the radical of such a ring R, then any R/J-module, e.g., J/J^2, is semisimple. In contrast, the existence of so many types of simple Lie algebras causes the proofs of many theorems on Lie algebras to devolve into seemingly countless special cases. Thus, while the variety of simple Lie algebras lends interest to the subject, the application of their structures makes the ensuing mathematics more boring! The same observation can be made about the structure theory of simple rings inasmuch as there are many seemingly different, i.e., nonisomorphic types, and, as we have shown, module theory over them is quite complex even for the most explicit types of simple rings.

The magnitude of the task of classifying simple rings is the *raison d'être* of this volume. While primitive rings and prime rings have evolved as fruitful generalizations of simple rings, and (in the latter case) of integral domains, there are just too many to classify. At present, the determination of the structure of simple rings with ascending chain conditions offers challenges enough to occupy the talents of mathematicians in the foreseeable future. (See Open Problems, Chapter 7.)

Appendix

1. Homological dimension

If $M \in \text{mod-}R$, a **projective resolution** for M, $P(M)$, is an exact sequence

$$\ldots \to P_n \to P_{n-1} \to \ldots \to P_1 \to P_0 \to M \to 0,$$

where each P_j is projective. Clearly, each module has a projective resolution. If $P_i = 0$, $\forall i \geq n + 1$, we say that $P(M)$ has length n. The greatest lower bound of the lengths over all projective resolutions for M_R is called the **projective** or **homological dimension of** M_R and is denoted pd M_R, pd M or hd M. By a trivial extension of Schanuel's lemma, one can show that if pd $M = n$ and

$$0 \to K \to P_{n-1} \to \ldots \to P_0 \to M \to 0$$

is exact with P_i projective for all i then K is projective. $\sup\{\text{pd } M \mid M \in \text{mod-}R\}$ is called the **right global dimension of** R and will be denoted r.gl.dim R. The Auslander global dimension theorem reduces the computation of r.gl.dim R to that of the projective dimensions of the right ideals.

THEOREM (Auslander) *If R is not a semisimple ring, then*

$$\text{r.gl.dim } R = \sup\{\text{pd } I \mid I \text{ a right ideal}\} + 1.$$

If M and C are any two R-modules and $0 \to K \xrightarrow{f} P \xrightarrow{g} M \to 0$ is exact with P projective, then

$$0 \to \hom_R(M, C) \xrightarrow{g^*} \hom_R(P, C) \xrightarrow{f^*} \hom_R(K, C) \to \operatorname{cok} f^* \to 0$$

is exact. $\operatorname{cok} f^*$ is usually denoted $\operatorname{Ext}_R{}^1(M, C)$. $\operatorname{Ext}_R{}^1(M, C)$ is a functor: mod-$R \rightsquigarrow Ab$, contravariant in the first variable and covariant in the second. Moreover $\operatorname{Ext}_R{}^1(M, C) = 0$ $\forall C (\forall M)$ iff M is projective (C is injective). (See Lambek [66] or Cartan–Eilenberg [56].)

Flat Modules and Regular Rings

PROPOSITION AND DEFINITION *A module* M_R *is called* **flat** *if it satisfies any of the following equivalent properties:*

(1) $M \otimes_R (-)$: R-mod $\rightsquigarrow$ *Ab is an exact functor;*

(2) *for all (finitely generated) left ideals of* R,

$$M \otimes I \overset{\text{can}}{\approx} MI$$

where the canonical map is the one induced by $M \otimes I \to M \otimes R = M$.

PROPOSITION AND DEFINITION *R is called* **(Von Neumann) regular** *if R satisfies any of the following equivalent properties:*

(1) *each* $M \in$ mod-R *is flat;*

(2) *each finitely generated right ideal of R is generated by an idempotent;*

(3) $\forall a \in R$, $\exists x \in R$ *such that* $axa = a$.

UTUMI'S THEOREM (Utumi [56]) *Let* M_R *be injective,* $\Lambda = \text{End}\, M_R$ *and* N, *the ideal of* Λ *consisting of all* $\lambda \in \Lambda$ *satisfying* $M \nabla \ker \lambda$. *Then*

(1) Λ/N *is regular;*

(2) $N = \text{rad}\, \Lambda$.

2. Categories, functors and natural equivalences

A category G consists of a class of **objects** obj G; for each ordered pair of objects, (A, B), a set $\text{Mor}_G(A, B)$, called the **morphisms** or **maps** from A to B; and for each ordered triple of objects, (A, B, C), a map (called **composition**)

$$\text{Mor}_G(B, C) \times \text{Mor}_G(A, B) \to \text{Mor}_G(A, C)$$

satisfying

(1) $\text{Mor}_G(A, B) \cap \text{Mor}_G(A', B') \neq \varnothing$ iff $A = A'$ and $B = B'$;

(2) each $f \in \text{Mor}_G(A, B)$ has a left identity in $\text{Mor}_G(B, B)$ and a right identity in $\text{Mor}_G(A, A)$;

(3) composition is associative.

Example mod-R where objects are right R-modules and maps, R-linear homomorphisms.

A map f from A to B will be denoted by the standard symbol $f: A \to B$.

A map $f: A \to B$ is called an **equivalence** if there exists $g: B \to A$ such that $fg = 1_B$ and $gf = 1_A$. (Of course, fg is the composition of f and g and 1_B is the (necessarily unique) left identity for f).

Let G and G' be categories. A **(contravariant) covariant functor** $T: G \rightsquigarrow G'$ is a rule associating with each $A \in G$, an object $T(A) \in G'$, and for each map $f: A \to B$, a map $T(f): T(A) \to T(B)$. $(T(f): T(B) \to T(A))$ satisfying

(1) $T(1_A) = 1_{T(A)}, \forall A \in G$;

(2) $T(fg) = T(f)T(g), \forall f, g$,

$$(T(fg) = T(g)T(f), \forall f, g).$$

Example $\forall A \in G$, we have two functors

(1) $\mathrm{Mor}_G(A, -): G \rightsquigarrow$ SETS,

(2) $\mathrm{Mor}_G(-, A): G \rightsquigarrow$ SETS,

defined as $\hom_R(A, -)$ and $\hom_R(-, A)$ respectively. (1) is a covariant functor and (2) is a contravariant functor.

The unadorned term functor means covariant functor. Two functors $S, T: G \rightsquigarrow G'$ are said to be **naturally equivalent**, $S \simeq T$, if for each $A \in G$ there exists an equivalence $\eta_A: S(A) \to T(A)$ such that for all $f: A \to B$ the diagram

$$\begin{array}{ccc} S(A) & \xrightarrow{\eta_A} & T(A) \\ \downarrow{\scriptstyle S(f)} & & \downarrow{\scriptstyle T(f)} \\ S(B) & \xrightarrow{\eta_B} & T(B) \end{array}$$

commutes.

We call a covariant functor $T: G \rightsquigarrow G'$ an **equivalence** if there exists $S: G' \rightsquigarrow G$ such that $T \circ S \simeq 1_{G'}$ and $S \circ T \simeq 1_G$. In this case, we say that G and G' are **equivalent** and indicate this by $G \approx G'$.

Finally, a functor $T: G \rightsquigarrow G'$ is called a **left adjoint** for

$S : G' \rightsquigarrow G$ if $\forall A \in G$, $\mathrm{Mor}_{G'}(T(A),\ ) \simeq \mathrm{Mor}_G(A, S(\ ))$ and similarly for the other variable.

EXAMPLE If $G = \text{mod-}A$, $G' = \text{mod-}B$, and N, an (A, B)-bimodule, then $T = \otimes_A N$ is the left adjoint for $S = \hom_B(N,\ )$.

Notes

Chapters 1 and 2

The Introduction indicates a brief history of the subject of non-Artinian simple Noetherian rings, introduced by the Goldie and Lesieur–Croisot Theorems in 1958–60. The Noetherian chain conditions themselves were introduced by Noether in 1921 in carrying over results and techniques of algebraic number theory and algebraic geometry to commutative rings satisfying the ascending chain condition (thus, Noetherian rings). In 1927, Artin generalized some of the Wedderburn Theorems for finite dimensional algebras to noncommutative rings satisfying the (ascending and) descending chain conditions for right ideals. (Later, Hopkins and Levitzki, independently, in 1939, showed that the descending chain condition implies the ascending chain condition, that is, they proved that right Artinian rings are right Noetherian.)

In view of the historical aspect of the Introduction, and *ad hoc* references to key theorems in Chapters 1 and 2, further comments may seem uncalled for. Nevertheless, we have not mentioned the Popesco–Gabriel [64] Theorem which states that for any abelian category C with generator U and exact direct limits, there is a full embedding $\operatorname{Hom}_C(U, \ ) : C \rightsquigarrow \text{mod-}B$ where $B = \operatorname{End}_C U$, with exact left adjoint S. This very general theorem provides additional insight for the correspondence theorem for projective modules of Faith [72*a*] (see 1.13), one of the main tools used in obtaining the structure of simple Noetherian rings given in Chapter 2. To wit, the adjoint functor S induces a category equivalence $C \approx \text{mod-}B/K$ where $K = \ker S = \{X \mid S(X) = 0\}$, and mod-$B/K$ is the Gabriel–Grothendieck quotient category (defined in Gabriel [62]). Thus, C is a 'localization' of mod-B in the sense of Gabriel [62], or a 'torsion theory' of mod-B in the sense of Lambek [71].

It is clear that this can be applied to a simple ring A with uniform right ideal U, since U is a generator of C = mod-A. In this case, as we have indicated in the Introduction and elsewhere, $B = \text{End}_C U$ is a right Ore domain, and thus the category mod-A is a quotient category of mod-B. Moreover, the functor $\text{Hom}_C(U, \ ): C \rightsquigarrow$ mod-B is a full embedding.

Chapter 3

The best known example of a non-Artinian simple Noetherian domain is what is called the 'algebra of quantum mechanics' or the Jordan–Weyl–Littlewood algebra A_1. The first systematic study of this algebra which we have found appears in Littlewood [33]. Here, a laborious construction of a quotient field for A_1 is given, anticipating Ore's more general results. Independently, Ore [31, 33] introduced the general ring of skew polynomials $k[t; \rho, \delta]$, established its Euclidean-like properties, and gave the general construction of a classical quotient field for an arbitrary Ore domain. Subsequently, Hirsch [37] generalized some of the Littlewood algebras to $2n$ generators, providing necessary and sufficient conditions for their simplicity.

Twenty-five years later, Rinehart [62] showed that r.gl.dim A_1 = 1 and in general, r.gl.dim $A_n \leq 2n - 1$ whenever the characteristic of k equals 0; r.gl.dim $A_n = 2n$ otherwise. Here, A_n is the ring of linear differential operators in $\partial/\partial X_1, \ldots, \partial/\partial X_n$, over the partial differential ring $k[X_1, \ldots, X_n]$. Roos [72] has shown that r.gl.dim $A_n = n$ if the characteristic of k equals 0, settling a special case of a problem posed in Rinehart's paper.

Rinehart also showed that these rings can arise in a different algebraic setting, namely, as certain quotients of the universal enveloping algebra of a finite dimensional abelian Lie algebra. In the same direction and attesting to the ubiquity of simple Noetherian rings, Dixmier [63] has shown that if g is a finite dimensional nilpotent Lie algebra over an uncountable, algebraically closed field of characteristic 0, and I is an ideal of the universal enveloping algebra A of g, then $A/I \approx A_m$ for some $m > 0$, if and only if I is maximal (see also Cohn [61*b*]).

The necessity of 3.2 (for the case of simple R) appears in Amitsur [57]; the necessity of 3.2(a) is implicit in Hart [71]. Also 3.3 – 3.5 are the Krull dimension 1 case of more general results due to Hart (same paper). 3.6(d)–(f) are due to Webber [70]. Finally, the generalities on differential modules are due to J. L. Johnson [71].

Chapter 4

The Goldie and Faith–Utumi Theorems place any simple Noetherian ring R between F_n and D_n where F is a right Ore domain with right quotient field D, and n an integer ≥ 1. This matrix representation has the following drawback: R may properly contain F_n, since in general, F is only a ring-1. The precise relationship between R and F is unknown. For example, does there exist a maximal order F of D such that $F_n \hookrightarrow R$? And if so, are any two of them equivalent right orders?

Let M denote the $n \times n$ matrix units of D_n, let $A = \{r \in R \mid rM \subseteq R\}$, and $B = \{r \in R \mid Mr \subseteq R\}$. Then, as we have shown in the Faith–Utumi theorem, $U = B \cap D = A \cap D$ is an ideal of $P = R \cap D$, and P_n, U_n, $U_n{}^2$ are right orders equivalent to R. Moreover, the proof shows that P, U, and U^2 are equivalent right orders of D.

Thus far we have not needed the assumption that R is simple, that is, the above holds for any right order R in D_n. However, when R is simple, then a theorem of Faith [65] states that B, A, BA, and U^2 are all simple rings, that is, F in the Faith–Utumi theorem can be chosen to be a simple ring-1. In this case it is easy to see that R is then right Noetherian iff F satisfies the a.c.c. on idempotent right ideals. It would be interesting to know if F can be chosen to be unique up to equivalence of orders.

In representing a simple Noetherian ring $A = \text{End}_B U$, the cut down lemma enables one to choose an Ore domain B with at most 1 nontrivial ideal but then, unlike A, B may not be a maximal order in its quotient ring. On the other hand, by the maximal order theorem, we can choose U and B so that B is a maximal order but then B, unlike A, may have an infinite ideal lattice. What properties other than maximality which are inherited when A is simple and Noetherian, is an interesting question which deserves further exploration.

We should also point out that the results in the section on maximal orders bear striking similarity to those for maximal orders in the classical sense (see Auslander–Goldman [60] and Ramras [69], [71]).

Chapters 5 and 6

Michler–Villamayor [73] have characterized right V-rings having right Krull dimension ≤ 1 (that is, right V-rings R such that R/I is Artinian for all essential right ideals I) as rings similar to finite products of rings, each of which is a matrix algebra over a right Noetherian PCI-domain. (The theorem is also a consequence of techniques used repeatedly in Chapters 5 and 6.)

Goodearl [72] showed that if every proper cyclic right R-module is semisimple, then R is right hereditary and a finite product of simple rings (etc.), reminiscent of the original constructions of Cozzens [70].

The observation that the solvability of systems of linear differential equations and the extensions of the base field are intimately connected, is due to Malgrange [66]. The present discussion, as well as (1) $\Rightarrow$ (2) of 5.21 and its elegant proof, are due to J. Johnson. (The original proof of the remaining implications given in Cozzens [70] is more complicated.)

q-fields were invented by Osofsky to produce examples of V-domains (PCI-domains) with infinitely many nonisomorphic simples (see Osofsky [71]). The example following 5.26 is taken directly from that paper.

Most of Chapter 6 is taken from Faith [73*b*], and generalizes the results of Boyle [74] (see 6.26) to non-Noetherian PCI-rings.

References

Albert, A. A. (1939) *Structure of Algebras*, Amer. Math. Soc. Colloq. Publ., Vol. 24, Providence, R.I.

Amitsur, S. A. (1957) Derivations in Simple rings, *Proc. Lond. Math. Soc.* (3), **7**, 87–112.

Amitsur, S. A. (1972) On central division algebras, *Israel J. Math.* **12**, 408–20.

Artin, E. (1927) Zür Theorie der hyperkomplexen Zahlen, *Abh. Math. Sem. Univ. Hamburg* **5**, 251–60.

Auslander, M. & Goldman, O. (1960) Maximal orders, *Trans. Amer. Math. Soc.* **97** (1), 1–24.

Bass, H. (1960) Finitistic dimension and a homological generalization of semi-primary rings, *Trans. Amer. Math. Soc.* **95**, 466–88.

Bass, H. (1962) *The Morita Theorems*, Mathematics Department, University of Oregon, Eugene.

Boyle, A. K. (1974) Hereditary QI-rings, *Trans. Amer. Math. Soc.* **192**, 115–20.

Brown, B. & McCoy, N. H. (1950) The maximal regular ideal of a ring, *Proc. Amer. Math. Soc.* **1**, 165–71.

Bumby, R. (1965) Modules which are isomorphic to submodules of each other, *Arch. Math.* **16**, 184–5.

Camillo, V. & Cozzens, J. H. (1973) A theorem on Noetherian hereditary rings, *Pacific J. Math.* **45**, 35–41.

Cartan, H. & Eilenberg, S. (1956) *Homological Algebra*, Princeton Univ. Press, Princeton.

Chatters, A. W. (1971) The restricted minimum condition in Noetherian hereditary rings, *J. Lond. Math. Soc.* (2) **4**, 83–7.

Chevalley, C. (1936) L'arithmetique dans les algebres de matrices, *Actualites Sci. Indust.*, No. 323, Paris.

Cohn, P. M. (1961*a*) Quadratic extensions of skew fields, *Proc. Lond. Math. Soc.* (3), **11**, 531–56.

Cohn, P. M. (1961*b*) On the imbedding of rings in skew fields, *Proc. Lond. Math. Soc.* (3), **11**, 511–30.

Cozzens, J. H. (1970) Homological properties of the ring of differential polynomials, *Bull. Amer. Math. Soc.* **76**, 75–9.

Cozzens, J. H. (1972) Simple principal left ideal domains, *J. Algebra* **23** (1) 66–75.

Cozzens, J. H. (1975) Reflexive modules and maximal orders, preprint.

Cozzens, J. H. & Johnson, J. L. (1972) An application of differential algebra to ring theory, *Proc. Amer. Math. Soc.* **31**, 354–6.

Deuring, M. (1948) *Algebren*, Chelsea, New York.

Dixmier, J. (1963) Représentations irreductibles des algèbres de Lie nilpotentes, *Anais Acad. Brasileira de Ciencias* **35**, 491–519.

Eisenbud, D. & Robson, J. C. (1970*a*) Modules over Dedekind prime rings, *J. Algebra*, **16**, 67–85.

Eisenbud, D. & Robson, J. C. (1970*b*) Hereditary Noetherian prime rings, *J. Algebra*, **16**, 86–104.

Faith, C. (1964) Noetherian simple rings, *Bull. Amer. Math. Soc.* **70**, 730–1.

Faith, C. (1965) Orders in simple Artinian rings, *Trans. Amer. Math. Soc.* **114**, 61–5.

Faith, C. (1966) Rings with ascending chain condition on annihilators, *Nagoya Math. J.* **27** (1), 179–91.

Faith, C. (1967*a*) *Lectures on Injective Modules and Quotient Rings*, Lecture Notes in Mathematics, Springer-Verlag, New York–Berlin.

Faith, C. (1967*b*) A general Wedderburn theorem, *Bull. Amer. Math. Soc.* **73**, 65–7.

Faith, C. (1971) Orders in semilocal rings, *Bull. Amer. Math. Soc.* **77**, 960–2.

Faith, C. (1972*a*) A correspondence theorem for projective modules, and the structure of simple Noetherian rings (*in* Proceedings of the Conference on Associative Algebras, Nov. 1970), *Institute Nazionale di Alta Matematica, Symposium Matematica*, **8**, 309–45.

Faith, C. (1972*b*) Modules finite over endomorphism ring, *Proceedings of the Tulane University Symposium in Ring Theory, New Orleans, 1971*, Lecture Notes in Mathematics No. 246, pp. 145–90, Springer-Verlag, New York–Berlin.

Faith, C. (1973*a*) *Algebra: Rings, Modules and Categories*, Springer-Verlag, New York–Berlin.

Faith, C. (1973*b*) When are cyclics injective?, *Pacific J. Math.* **45**, 97–112.

Faith, C. (1974*a*) On the structure of indecomposable injective modules, *Comm. in Algebra*, **2**, 559–71.

Faith, C. (1974*b*) On hereditary rings and Ann Boyle's conjecture.

Faith, C. & Utumi, Y. (1965) On Noetherian prime rings, *Trans. Amer. Math. Soc.* **114**, 1084–9.

Faith, C. & Walker, E. (1967) Direct-sum representations of injective modules, *J. Algebra*, **5** (2), 203–21.

Findlay, G. & Lambek, J. (1958) A generalized ring of quotients I, II, *Can. Math. Bull.* **1**, 77–85, 155–67.

Gabriel, P. (1962) Des categories abeliennes, *Bull. Soc. Math. France*, **90**, 323–448.

Goldie, A. W. (1958) The structure of prime rings under the ascending chain condition, *Proc. Lond. Math. Soc.* (3), **8**, 589–608.

Goldie, A. W. (1960) Semiprime rings with maximum condition, *Proc. Lond. Math. Soc.* (3), **10**, 201–20.

Goldie, A. W. (1962) Noncommutative principal ideal rings, *Archiv. Math.* **13**, 213–21.

Goldie, A. W. (1969) Some aspects of ring theory, *Bull. Lond. Math. Soc.* **1**, 129–54.

Goodearl, K. R. (1972) Singular torsion and the splitting properties, *Amer. Math. Soc. Memoirs*, No. 124.

Goodearl, K. R. (1975) Subrings of idealizer rings, *J. Algebra*, **33** (3), 405–29.

Hart, R. (1967) Simple rings with uniform right ideals, *J. Lond. Math. Soc.* **42**, 614–7.

Hart, R. (1971) Krull dimension and global dimension of simple Ore-extensions, *Math. Zeit.* **121**, 341–5.

Hart, R. & Robson, J. C. (1970) Simple rings and rings Morita equivalent to Ore domains, *Proc. Lond. Math. Soc.* (3) **21**, 232–42.

Hirsch, K. A. (1937) A note on noncommutative polynomials, *J. Lond. Math. Soc.* **12**, 264–6.

Jacobson, N. L. (1943) *The Theory of Rings*, Amer. Math. Soc. Mathematical Surveys, Vol. 2, Providence, R.I.

Jacobson, N. (1945) The structure of simple rings without finiteness assumptions, *Trans. Amer. Math. Soc.* **57**, 228–45.

Jacobson, N. L. (1964*a*) *Structure of Rings*, rev. ed., Amer. Math. Soc. Colloq. Publ., vol. **37**, Providence, R.I.

Jacobson, N. L. (1964*b*) *Lectures in Abstract Algebra*, Vol. **3**, *The Theory of Fields*, Van Nostrand, New York.

Jans, J. P. (1964) *Rings and Homology*, Holt, New York.

Jategaonkar, A. V. (1969) Ore domains and free algebras, *Bull. Lond. Math. Soc.* **1**, 45–6.

Jategaonkar, A. V. (1971) Endomorphism rings of torsionless modules, *Trans. Amer. Math. Soc.* **161**, 457–66.

Johnson, J. L. (1971) Extensions of differential modules over formal power series rings, *Amer. J. Math.* **43** (3), 731–41.

Johnson, R. E. (1951) The extended centralizer of a ring over a module, *Proc. Amer. Math. Soc.* **2**, 891–5.

Johnson, R. E. & Wong, E. T. (1959) Self-injective rings, *Can. Math. Bull.* **2**, 167–73.

Johnson, R. E. & Wong, E. T. (1961) Quasi-injective modules and irreducible rings, *J. Lond. Math. Soc.* **36**, 260–8.

Kaplansky, I. (1952) Modules over Dedeking rings and valuation rings, *Trans. Amer. Math. Soc.* **72**, 327–40.

Köthe, G. (1935) Verallgemeinerte Abelsche Gruppen mit Hypercomplexen operatoren ring, *Math. Zeit.* **39**, 31–44.

Lambek, J. (1966) *Lectures on Rings and Modules*, Ginn-Blaisdell, Waltham, Mass.

Lambek, J. (1971) *Torsion Theories, Additive Semantics, and Rings of Quotients*, Lecture Notes in Mathematics, No. 177, Springer-Verlag, New York–Berlin.

Lesieur, L. & Croisot, R. (1959) Sur les anneaux premiers noetheriens a gauche, *Ann. Sci. Ecole Norm. Sup.* **76**, 161–83.

Levy, L. (1963) Torsion free and divisible modules over nonintegral domains, *Can. J. Math.* **15**, 132–57.

Littlewood, D. E. (1933) On the classification of algebras, *Proc. Lond. Math. Soc.* (2), **35**, 200–40.

Malgrange, B. (1966) *Cohomologie de Spencer* (*d'apres Quillen*), Orsay.

Matlis, E. (1958) Injective modules over Noetherian rings, *Pacific J. Math.* **8**, 511–28.

Matlis, E. (1968) Reflexive domains, *J. Algebra*, **8**, 1–33.

Michler, G. (1969) On quasi local Noetherian rings, *Proc. Amer. Math. Soc.* **20**, 222–4.

Michler, G. O. & Villamayor, O. E. (1973) On rings whose simple modules are injective, *J. Algebra*, **25**, 185–201.

Morita, K. (1958) Duality for modules and its applications to the theory of rings with minimum condition, *Sci. Rep. Tokyo Kyoiku Daigaku Sect. A*, **6**, 83–142.

Ore, O. (1931) Linear equations in non-commutative fields, *Ann. Math.* 32, 463–77.

Ore, O. (1933) Theory of non-commutative polynomials, *Ann. Math.* **34**, 480–508.

Ornstein, A. Ph.D. Thesis, Rutgers University, 1967.

Osofsky, B. L. (1968) Noninjective cyclic modules, *Proc. Amer. Math. Soc.* **19**, 1383–4.

Osofsky, B. L. (1971) On twisted polynomial rings, *J. Algebra*, **18**, 597–607.

Papp, Z. (1959) On algebraically closed modules, *Publ. Math. Debrecen.* **6**, 311–27.

Popesco, N. & Gabriel, P. (1964) Caractérisations des catégories abeliennes avec générateurs et limites inductives exactes, *C.R. Acad. Sci. Paris*, **258**, 4188–90.

Ramras, M. (1969) Maximal orders over regular local rings of dimension two, *Trans. Amer. Math. Soc.* **142**, 457–79.

Ramras, M. (1971) Maximal orders over regular local rings, *Trans. Amer. Math. Soc.* **155** (2), 345–52.

Rinehart, G. S. (1962) Note on the global dimension of a certain ring, *Proc. Amer. Math. Soc.* **13**, 341–6.

Robinson, A. (1963) *Introduction to Model Theory and the Meta-Mathematics of Algebra*, North Holland, Amsterdam.

Robson, J. C. (1972) Idealizers and hereditary Noetherian prime rings, *J. Algebra*, **22**, 45–81.

Roos, J. E. (1972) Algébre homologique. Determination de la dimension homologique globale des algébres de Weyl, *C. R. Acad. Sci. Paris Ser. A*, **274**, 23–6.

Sandomierski, F. L. (1968) Nonsingular rings, *Proc. Amer. Math. Soc.* **19**, 225–230.

Silver, L. (1967) Noncommutative localizations and applications, *J. Algebra*, **7**, 44–76.

Steinitz, E. (1948) *Algebraische Theorie der Korper*, Chelsea, New York.

Swan, R. G. (1962) Projective modules over group rings and maximal orders, *Ann. Math.* **76**, 55–61.

Utumi, Y. (1956) On quotient rings, *Osaka Math. J.* **8**, 1–18.

Utumi, Y. (1963) A Theorem of Levitzki, *Math. Assoc. of Amer. Monthly*, **70**, 286.

Webber, D. B. (1970) Ideals and modules of simple Noetherian hereditary rings, *J. Algebra*, **16**, 239–42.

Wedderburn, J. H. M. (1908) On hypercomplex numbers, *Proc. Lond. Math. Soc.* (2), **6**, 77–117.

Zelmanowitz, J. M. (1967) Endomorphism rings of torsionless modules, *J. Algebra*, **5**, 325–41.

Index